This I
the la

THE LEARNING CENTRE
TOWER HAMLETS COLLEGE
ARBOUR SQUARE
LONDON E1 0PS

RESEARCHING LANGUAGE

Working with English Language Data at AS/A Level and Beyond

Heinemann Educational Publishers
Halley Court, Jordan Hill, Oxford OX2 8EJ
A division of Reed Educational & Professional Publishing Ltd
Heinemann is a registered trademark of Reed Educational & Professional Publishing Ltd

OXFORD MELBOURNE AUCKLAND
JOHANNESBURG BLANTYRE GABORONE
IBADAN PORTSMOUTH NH (USA) CHICAGO

First published 1993
Reprinted 1997
Reprinted 1998
Second edition 2000

ISBN 0435 13270 9

2004 2003 2002 2001 2000
10 9 8 7 6 5 4 3 2 1

Editor: Karen Westall
Layout artist: TechType, Abingdon, Oxon
Illustrations: Lynne Dougall and Kim Woolley

Acknowledgements
The author and publishers wish to thank the following for permission to reprint copyright material:

BT Cellnet for use of the photograph (p. 6); *The Guardian* for 'Black Wash' by Frank Martin, 15 August, 1984 (p. 17); Edinburgh University Press for 'Inventory' by Liz Lochhead in *Dreaming Frankstein and Collected Poems* (Polygon) (p. 30); The NSPCC for their advertisement (p. 31); Labour's Education Action Zones (p. 40); The Labour Party for Tony Blair's speech on the night of the general election 1997 (pp. 63–64); Granada Television for the extract from *Coronation Street* 20 July, 1998 (pp. 69–70); *The Guardian* for 'Fuss about "fcuk" ads' by Kamal Ahmed, 21 January, 1998 (p. 79); R. and W. Heap (Publishing) Company Ltd. for 'Shamed By Your Mistakes In English?' advertisement (p. 79); *The Guardian* for 'Sprechen Sie German?' by Ian Traynor, 5 February, 1998 (p. 80); Carlton Cards for use of the greetings cards (pp. 102–103); The Society of Authors as the literary representative of the Estate of Virginia Woolf for *A Haunted House* (pp. 119–120); PA News Centre for the reports on the Shipman trial (pp. 126–128); *Glossop Advertiser* for 'Christmas ghost in white returns to haunt Hillmen' 1 January, 2000 (p. 131); *Woman* magazine for problem pages (pp. 136–138); D. C. Thomson & Co. Ltd. for an extract from *The Beano*, 1998 (p. 148); Reproduction of the OVALTINE advertisements (pp. 152–156) with kind permission of Novartis, legal successor of Wander UK; Whipsnade Wild Animal Park for the penguin photograph (p. 158); The Met. Office for the two weather forecasts (pp. 190–191).

The publishers have made every effort to trace the copyright holders, but if they have inadvertently overlooked any, they will be pleased to make the necessary arrangements at the first opportunity.

Printed and bound in the UK, by Bath Press

The Author

Angela Goddard has taught English across all phases of education, from primary to university level. She ran the Language in the National Curriculum Project (LINC) for Tameside, Stockport and Manchester, 1990–1992. Currently she is Senior Lecturer in Linguistics at the Centre for Human Communication, Manchester Metropolitan University. From 1983 to 1995 she was Chief Moderator for NEAB's English Language A Level Investigations, and is now Chair of Examiners for English Language AS/A Level at AQA. She is co-editor (with Professor Ronald Carter) of the Routledge *Intertext* series.

Acknowledgements

Grateful thanks to all the English Language A level students who have contributed data and ideas to this book. The contributions of the following groups and individuals are also gratefully acknowledged:

Dr Alan Cruttenden, Manchester University Dept. of Linguistics, for material on intonation; students and staff from Gosforth High School, Newcastle Upon Tyne, particularly Sarah Haggart, for the answerphone data; Ram Wallace, for the Newcastle United match write-up; Adrian Beard, for the Tony Blair speech; LINC Working Party members: in particular, Liz Armstrong for the material on weddings; Tony Meheran for the material on the language of gravestones; Jane Hamshere and Carol Ashton for the 'birth cards' idea; Fiona Stiles, Janet Lee and Karen Moorcroft for the different versions of 'The Argument'; The LINC Primary Advisory Teachers for Tameside, Stockport and Manchester, and Working Party Members: in particular, Ann Howard for transcripts of young children talking; Jan Turner for the nursery writing; Judith Chapman for 'The Tooth Fairy'; Sally Heap, English Primary Advisory Teacher for Stockport, for 'Little Bonkey'; Rob Greenall, Manchester Metropolitan University, for 'Incey Wincey Spider'.

RESEARCHING LANGUAGE

Working with English Language Data at AS/A Level and Beyond

Angela Goddard

Contents

Introduction

Courses change considerably over time. Even the term 'courses' is no longer universally applicable: if you look across the different phases of education, you will find such varied terms as 'programmes of study', 'specifications', 'modules' and 'units', to name but a few.

One change that can be observed particularly at AS/A level and in Higher Education is the move towards modularisation. As part of this process, students are being increasingly asked to show that they can be flexible, study a range of areas and work independently, managing their own learning. Demands on staff are often about how to give students enough of a foundation to enable them to be independent and generate their own culture of enquiry. In a number of AS/A level specifications, there are formal assessment points for investigative work, consisting of a piece of coursework research or an examined equivalent. However, investigative *skills* underpin all English Language AS/A level work, in the same way that they form the foundation for courses at higher levels of study. Investigative skills are research by another name, and research is the route to academic development and success.

Best practice in language work in our schools and colleges involves as early focus as possible on how to work with language data, allowing students to develop their skills and knowledge by encountering material that raises interesting and varied questions about language. The principal aim of this book is to exemplify and endorse that good practice by providing material and approaches that have been seen, through extensive trialling and use, to encourage active learning from the beginning.

Possible ways of working are varied and flexible: for example, the book can form the basis of shared group work in classroom situations, with further research being undertaken independently by groups or individuals; students who are in the position of having to work alone, with only the occasional opportunity to meet with a tutor, will also find the book useful distance learning material which teaches necessary analytical skills and provides research ideas and approaches.

Researching Language has been reprinted and updated several times since it first appeared in response to the then need for accessible language material for students at A level. Since then, it has been used at pre- and post-A level, as well as at A level itself: for example, for secondary school National Curriculum Language Study work, and within Higher Education as a basic introductory text for non-specialist students. Its wide applicability shows that interesting language data can be analysed at many different levels, and that, while individual courses come and go, the principle of taking an investigative approach has a universal and timeless relevance.

Working instructions

The *Activities* provide a starting point for tackling data in particular research areas and then develop analysis in more detail by:

- encouraging reflection on findings, guiding students towards main points or issues and relating language features to the wider social contexts of use

- providing additional material which is designed to deepen and broaden the research outlook.

Research pathways give suggestions for further pieces of group or independent research.

Section A

LAYING THE FOUNDATIONS

1 Developing a research culture

Developing a good working atmosphere for language study is all about getting individuals to adopt a research approach to the language environment around them. The next unit will look specifically at the personal resources available to us as individuals; this unit provides a number of ways in which individuals can set up good research *habits* for themselves and, in so doing, contribute to the research culture of the larger group.

ACTIVITY

Read through the following list of suggestions and make a note of any that you already carry out on a regular basis. Where these ideas are new to you, think of ways in which you could put them into practice. Finally, add any further ideas of your own to the list.

1 Recording

Keep a note of any interesting examples of language use that you see and/or hear as you go about your everyday life. You've probably had that experience where you would have liked to have a tape-recorder going at the very moment someone was saying something; but this isn't an all-or-nothing situation, since you can also write down some of the words, phrases and routines you're hearing. Also, there are likely to be many examples of written language you observe on your travels. Sometimes, we think we will remember examples we have seen, but we invariably don't. For example, by looking through one of my notebooks, I rediscovered an example of a shop name I saw in Spain two years ago (Pamplona, 9 October, 1998). The name made me laugh a lot at the time, but I'd completely forgotten about it until I reread my notebook. The shop was a man's tie outlet called 'The Throttleman'. When I explained to my Spanish friends that this sounded like the sort of shop the Boston Strangler would have liked, they didn't know what I was talking about. That is, they'd heard of the Boston Strangler, but they didn't know what the word 'throttle' meant, so they'd taken it to connote some kind of fashion statement: instead of 'Armani Man', read 'Throttleman'.

Sometimes, written texts are impossible to collect because of their size. In this situation, you may want to make a photographic record instead of, or in addition to, a notebook entry. For example, overleaf is a recent billboard hoarding I photographed because of its play on the different meanings of 'PC'.

Billboard on Hyde Road, Manchester

Recording is also important in those situations where texts have a high turnover rate. For example, Internet texts are frequently changed. If you are interested in doing some work on an Internet text, you will need to save it to disk in order to ask permission of the authors for its use.

2 Referencing

Referencing is all about keeping careful notes on sources. The rationale for referencing is that others should be able to follow in your footsteps and find all the sources of information you are using as evidence in your argument or presentation of an idea.

References come in many different forms and are used for a variety of purposes. In the Internet example above, you would need the URL ('unique reference locator', or Internet address) and the date you downloaded your material, in order to compile a reference list for an essay or piece of research; but you would also need this for yourself anyway, to find the page again. It's surprising how many people forget to write down the addresses of sites they've visited.

If you are recording material in a notebook or photographing something, you need to note place and time details (see my example above). Other common resources, such as newspaper clippings (see below) also need the date and the name of the newspaper. Not only is this important for academic purposes, but for practical ones, too: if you want to write to the newspaper about the material you've collected, they won't be able to locate the material in their files without a date; and if you don't know which paper your material is from, you won't be able to write to them at all.

3 Public resources

As has already been suggested above, there are many public resources that can supply rich data for language study. The Internet is of course an important resource which, aside from the information that can be accessed, is in itself a wonderful language environment.

As you go through the book, you will be given many different Internet language pages to check out. As you will be aware, such pages change frequently, so be prepared to use the links that are here but also be alert for new pages that you can share with others in your group.

Aside from Internet texts, there are of course many traditional paper-based texts that can be scrutinised for language data. For example, daily newspapers often contain articles and readers' letters about language as a topic – for example, on issues such as the following:

- 'Political correctness' in language
- Attitudes to accents and dialects
- The invention of new terms
- New communication tools
- Language use in advertising
- Swearing in films and TV programmes
- New archives (such as dictionaries and grammars) that show changed rules for 'correctness'
- Significant themes from academic conferences on language
- Language policies in schools
- Language policies in different countries

Textbooks and journals are, of course, a crucial part of a research culture.

4 Reading, note-making and presentations

One way to work collaboratively is to organise sessions where students take it in turn to read some material and present some notes on it. This is a good way to train yourself to read with a purpose, and to select important points to present to others. It can also be a point where you decide you want to be tested on key skills, such as your use of IT or your ability to communicate in group contexts. If you present some quantitative survey work, you could even cover some numeracy skills at the same time.

Don't limit the idea of presentation just to conveying your ideas from your reading: summarising and speculating on any of the material that has been described above could also constitute presentation work of an interesting nature.

5 Filing

This is no one's favourite topic, but it's very important. You need to find a way to organise all the material that has been discussed in this unit, if you are to make the most of your collections and analyses. Obviously, much will depend on the type of module you are engaged in, since there will be headings there that you will need to work to. But, as a general principle, allocate some separate notebooks, folders and box files for the different aspects of language you are covering.

2 Reviewing your credentials

The following questionnaire is designed to get you thinking about some of the *personal* language resources you have around you.

ACTIVITY

Complete the questionnaire below. You may find it useful to work in pairs, so that you can jog each other's memory about resources you might overlook.

If it's appropriate, report back briefly to the group on one or two particular aspects you are interested in and/or have resources for. In reporting back, you may find that people have some interesting material that might be added to that of others and used in comparative ways — for example, old comics aimed at girls, compared with those aimed at boys; magazines aimed at men compared with those aimed at women.

QUESTIONNAIRE

(i) *Your own language development*

Make a list of everything you still have in your possession that relates to your own language development. Here are some examples:

School exercise books   Old comics or magazines
Early readers   Tape-recordings of you
Teachers' reports   Old diaries, scrapbooks or other
                                 notebooks

(ii) *Your family*

Write down some details in note form in response to the following questions:

- Do you have younger relatives who are at the early stages of learning language?
- Do the people in your immediate family use a particular accent, dialect, or know more than one language?
- Is there anyone in your family who has had particular language problems?

continued

- Do you have relatives who work in occupations that use a specific type of language?
- Do the older members of your family have any collections of particular types of material (e.g. football programmes, greeting cards)?
- Do you think you use language differently from your older relatives? If so, how?
- Do particular family members fall into certain 'roles' in group conversations (e.g. peacemaker, agitator, listener)?

(iii) *You now*

- Do you belong to any groups or take part in any leisure activities that have their own forms of language?
- Do you do any part-time work that involves you in using language in a certain way?
- Do you collect, read and/or write particular sorts of material?
- Do any of your friends have particular accents and dialects, or use more than one language? (e.g. Do you have any foreign penfriends?)

(iv) *And your addictions*

- What kind of an addict are you?
- What are some things that fascinate you about the language you see and hear around you?
- For example, are you interested in any of the types of language set out below?

Choose *three* items from the list, or make up your own list of three types of language that particularly intrigue you, and try to decide what it is about them that fascinates you.

The conversations people have on buses or trains	People's names
The way people speak on the telephone or to answering machines	Advertising tricks
	Persuasive speeches
	Body language
	Certain types of literature
Problem pages in magazines	Junk mail
Shop names	Car names
Road signs	Gravestones
Sign language	The way other languages work
Graffiti	Song lyrics
Ritual language used in ceremonies, e.g. weddings, funerals	Personal (small ads) columns
	Job jargon
	Bias in the Press
Greetings	The way animals communicate
Terms of address	Children's expressions

ACTIVITY

If you are working individually, make notes on your initial ideas, in preparation for your next meeting with your supervisor.

If you are working as a whole group, each person should choose one aspect of his or her notes to report back to the rest of the group. Set a deadline for all group members to bring in a piece of material that relates to some aspect of their notes. Each member of the group should then give a five-minute presentation of this material, explaining its source and what it indicates about the chosen aspect(s) of language.

3 What do you know?

As a language user, you already know a great deal of language. How can you convert what you know implicitly, as a language user, into conscious knowledge which can form the basis of an analysis? This unit is going to help you to answer that question.

Decoding written texts

Readers – even very young ones – bring a lot of knowledge of different aspects of language to the decoding of written texts.

Reading is not simply recognising individual words; it is understanding how texts are put together, how they work. Good readers have an awareness of a specific number of language areas or *levels*.

Language level I: Graphology

You will be studying each of these areas separately, but of course in reality they do not exist in isolation. As you move from one level to the next, try to carry forward what you have learned and to apply that learning to the new work you encounter.

Graphology – which literally means 'the study of marks' – refers to all the visual aspects of the written language. So aspects like layout, typeface, punctuation, spelling, abbreviations, images and other artwork such as logos would be included here.

Don't be confused by the fact that, outside the area of Linguistics, the term 'graphology' is used to refer narrowly to the study of handwriting.

ACTIVITY

Look at the differently shaped texts on page 12.

Imagine that the horizontal lines are lines of writing.

How do the different layouts give you an indication of how these texts should be read, and what they are?

(Suggested answers on page 70.)

continued

continued

When you have finished, go on to look at the data below.

You will see below four different fonts (typefaces) which are among the many available in modern word-processing packages. These are from Word 97 and, in order of appearance, they are: 'funstuff'; 'cut and paste'; 'curlz'; 'digiface wide'.

What *graphological* devices are being used in these fonts, and what meanings are being conveyed in each case? Are the connotations of the fonts (i.e. the ideas we associate with them) appropriate to the overall meanings of the messages? If not, what kinds of messages are more usually written using these fonts?

If you have access to Word, experiment with some of the fonts available: you will find them listed under 'format' and then 'font' at the top of your screen.

Please note that, because you have not paid your telephone bill, we have been forced to disconnect you.

We are sad to announce the sudden death of our colleague, Mr. Smith

We would be grateful if you could attend for interview at this establishment

WE ARE PLEASED TO ANNOUNCE THE BIRTH OF OUR DAUGHTER, LUCY

ACTIVITY

If you are working in groups, look through a range of magazines and newspapers to find 10–15 symbols and logos. These do not have to be from adverts – they could be from small ads pages, magazines features, etc. Cut them out and paste them onto a sheet.

Make sure you don't give your readers any clues about where they've come from (i.e. by including brand-names for adverts) and try not to make them too obvious.

Make a note yourselves of their source. Number the cuttings.

continued

When you have finished, swap your sheets around, so that each group is given another's cuttings.

Each group should then write down their associations for each symbol or logo – what does it remind them of? Even if they know where the cuttings are from, they should try to say what image, idea or feeling they are given by each cutting.

Compare all the results with the answers, pin up all the sheets for everyone to see and discuss the following questions:

- Can you find any patterns in how symbols and logos are used?

- Are certain types of symbol used in our culture to suggest particular ideas?

- If any members of the group have knowledge of other cultures or societies, are there any symbols or images used there to suggest different ideas?

If you are working individually, make a collection of different types of symbols and logos and write down the details as above. If possible, find six informants and ask them what associations they have for each symbol or logo.

Try to find patterns in how the symbols and logos are used and in what people's associations are.

When you have finished, keep all your material as a reminder of what this language level is all about.

Website

Computer-based communication has provided us with new ways to communicate in writing – for example, in chat-rooms and on bulletin boards. These new contexts have generated new symbols, often composed of punctuation marks, where people express their emotions in graphic form. These are called 'emoticons', and two examples are the 'smiley' ☺ and the 'frowney' ☹ .

To view a gallery of emoticons go to:

http://www.randomhouse.com/features/davebarry/emoticon.htm

Language level II: Phonology

Phonology refers to the study of sound.

It may seem strange to think of considering sound in written language, but writers know that readers use an 'inner ear' when reading texts; sound patterns may also correspond to visual (i.e. letter) patterns, traceable by the eye. Manipulating patterns of sound can make a text more memorable; the use of devices such as rhyme, alliteration, and the use of puns which rely on sound, or playing sound off against spelling, are the staple of advertising copywriters. Writers of 'serious' texts are unlikely to use sound patterning extensively, because this can lead to a sense of lightheartedness and playfulness – as in tongue twisters, rhyming jingles and jokes.

ACTIVITY

Look at the data below. What phonological devices are being used in these texts, and what is their purpose?

TABLOID HEADLINES

Randy Mandy Gets High on Brandy

(Story about a woman who was accused of indecency while drunk on an airplane)

Web War as Surf Costs are Slashed

(Story that a communications company is to offer free phone calls for Internet access)

Colly Hits Back with a Hat-Trick

(Footballer Stan Collymore scores three goals for Leicester after being criticised for his behaviour)

HAIRDRESSERS' SHOP NAMES

The Hairport
Hairazors
Fellaz Hair Design
Classic Cutz
Aries
Tint Inn
The Men's Den

COCKTAIL NAMES

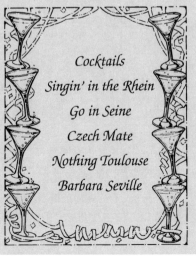

Cocktails

Singin' in the Rhein

Go in Seine

Czech Mate

Nothing Toulouse

Barbara Seville

ACTIVITY

Look through a range of tabloid newspapers for uses of phonological devices. If you are working in groups, try the following division of labour:

- One group should look for rhyme.

- One group should look for uses of sound symbolism (sometimes called 'onomatopoeia' – where the word imitates a sound).

- One group should look for vowel variation – for example, in the following headline, the vowel variation is as important as the alliteration on the letter 'b': big, bronze Buddha bagged by burglars.
- Two or more groups should look for alliteration (consonant repetition). These groups might focus on a different letter each.

Cut out all your examples.

If you find a feature which may be useful to another group, offer it to them for their collection.

When you have finished, paste your examples onto some display sheets, and pin them up for everyone to see. Invent a title for your display, then discuss the following questions:

- How far does the tabloid press play on sound to achieve certain effects in the language used?

- What effects are created, in your opinion?

If you are working individually, categorise your extracts, and try to answer the questions above.

Website

If you go to the website below and follow the instructions, you will arrive at a page of 'phonesthemes' which describes some of the associations we have for the different sounds of English:

http://www.onelook.com

Click on 'more dictionaries', then on 'dictionaries of a different kind'. You will then find a link called 'Dictionary of English Phonesthemes'.

Language level III: Semantics

This refers to the choice of words and phrases which, put together, weave a pattern of meaning. Many different aspects of language choice could be significant here, for example:

the level of formality – use of colloquial language, or abstract, Latin-based vocabulary (e.g. have a go/try/endeavour, kick-off/begin/commence)

the field of reference – terms that are all from the same area of knowledge or experience (e.g. medicine, computers, love, war) *connotation* – the way in which some words and phrases can evoke powerful association or feelings in the reader's mind, and make him or her view the subject matter in a certain light (e.g. freedom fighter/terrorist). Foreign terms are sometimes used to give particular impressions, playing on the stereotypes we have of different cultures (e.g. French 'romance' to sell perfume; German 'efficiency' to market cars)

the use of idiomatic language – language that works on more than the simply literal level (e.g. metaphor, pun). Idiomatic language can be put to a variety of uses; often it works towards constructing a particular viewpoint for the reader, sometimes by bringing different ideas together in a new way

collocation – the way in which certain items of language are expected to occur together, and in a certain order (e.g. fish and chips; he and she; Father, Son and Holy Ghost). Some collocations become clichés – rather worn-out and empty phrases (e.g. blushing bride; rack and ruin; come rain or shine)

language change – terms can die out, suggest a bygone age (be 'archaic'), be new coinages (neologisms), narrow or widen their meanings, or go up and down in respectability. Sensitivity to these changes is one of the ways we can date a text.

ACTIVITY

Look at pages 17–19. What *semantic* devices are being used in the photograph and these texts, and what meanings are conveyed by them?

GUARDIAN

Jubilant West Indies supporters spelt out their thoughts (above) after England's defeat yesterday while Joel Garner grabbed a couple of souvenirs of his six-wicket tally in the match.

© *The Guardian* (Frank Martin)

INNER SOUND AND VOICE WORKSHOPS

Among the Activities We Will Share:

- We will learn Mongolian overtone chanting, an ancient shamanic vocal practice which makes audible the natural harmonic spectrum of the voice in its pure rainbow colours, so that unearthly, angelic and bell-like tones are heard floating above the voices of the chanters.

- We will learn vocal purification practices and how to cleanse the chakras and subtle body.

- We will explore sonorous yogas and vocal practices involving the elements.

- We will explore, in order to forgive and clear, the patterns of our ancestral lines.

HAIRDRESSERS' SHOP NAMES

Power Cuts	Room at the Top
The Head Gardener	Headmasters
Blade Runners	A Head Start
As You Like It	Cuts Bothways

TOILET CISTERN NAMES

The Little Niagara
The Great Athenaeum
The Dauntless Bi-Flow
The Avalanche

NEWSPAPER HEADLINES

Nurses Upset by Cuts
Fireman Marries Old Flame
Mothers Who Smoke Have Lighter Children
Butter Price War Spreads

CAR NAMES

Punto, Clio, Micra
Ka, Polo, Escort
Fiesta, Civic, Proton
Boxer, Prelude, 306 XSi 2.0

CAFE LA MAISON

Specials
A tournedos of beef topped with a liver parfait,
enrobed in crepinette and oven baked,
served with a Madeira and truffle fondue
escorted by pommes sautes aux fines herbes

★ ★ ★ ★ ★ ★ ★ ★

continued

On my way out I knocked at Mr Contreras's door. He was inside again and much relieved to see me. I let his waves of information about the glazier wash over me, thanking him when there was a break in the surf, then explaining my going back.

Guardian Angel by Sara Paretsky

HERE
Lies in a horizontal position
the outside case of
THOMAS HINDE
Clock and watch maker,
Who departed this life
Wound up in hope
Of being taken in hand
By his Maker,
And being thoroughly cleaned,
Repaired and set a-going,
In the world to come,
On the 15th of August, 1830,
In the 19th Year of his Age.

Read through the penfriend's letter on page 20.

What aspects of English semantics are giving this writer some problems?

If you were working in small groups, feed back your results to the whole group. If you were working individually, make some written notes on your findings, as a summary for your own use.

Via N. Furnari n=° 45,
89100 Reggio CALABRIA
ITALIA

Dear Diane,

I also like you, I have been very busy, I am busy still. I am studing for University, I must study for two examinations for April and besides the Language and Literature English and Language and Literature Spanish for June. But the truth problem is the english language. I love it, but I have a professor very crazy. Your name is Bernard Dold, he is of Liverpool.

This year St. Valentine's day has been ugly because my boyfriend there isn't. He is making the military service. However also we use to send the cards to our boyfriends. Besides also we use presents, or a candlelight meal, or flowers.

I am happy because in short days will be my birthday: 31th March, I am 21. I am old! When is your birthday? Excuse me but I have forgotten it.

My boyfriend's name is Carmelo. I have been going-out with him five years.

. How are you? I hope well and I hope well also for your ill friends.   By for now

P.S. WRITE SOON, PLEASE—   love Maria

CIAO!

N.B.- I ~~don't~~ haven't understand the word "going-out".
It mean to go out, or to get engaged -?
. Please, write me correct form—.

ACTIVITY

Metaphor

Many of the everyday terms we use are metaphorical, but we use them so frequently that we often don't realise this. Look at the examples below:

I laughed my head off; he cried his eyes out; I nearly died laughing; button your lip; I'm in hurry – must fly!; her eyes were glued to the TV set; he broke my heart; I'm going out of my mind; pull yourself together; shake a leg; pull your socks up; our eyes met across the room; keep your nose out of my affairs ...

Sometimes it is possible to group a number of sayings together, as variations on one basic metaphor. For example:

Love as a Journey
We've come to the parting of the ways. You need to go your way, and I'll go mine.

We've come to a crossroads in our relationship. We've come a long way together, but this is the end of the road. Maybe in a while our paths will cross again ...

Argument as Warfare
She shot my argument down in flames. My argument was riddled with holes. We clashed, fought head-on, she attacked me and won. I had no back-up at all. She marshalled all her forces and ambushed me. I surrendered and we called a truce ...

Now take one of the metaphors below, and brainstorm all the terms you can think of, along the lines of *Love as a Journey* and *Argument as Warfare*:

A Lover as Food	Anger as a Boiling Liquid
Words as Weapons	Emotions as Colours
The Brain as a Machine	Time as Money
Anger as Fire	

If you have been working in groups, share your ideas with the whole group and discuss the following questions:

- Are there further metaphors you can think of, that are used in everyday conversations?

- Do you think that having metaphors such as the above conditions us to think in a certain way?

- What would be the difference in our way of thinking if we had the following metaphors?

    A Lover as a Building
    The Brain as a Flower
    Words as Smells

Try making up some more sayings, to see what these ideas would sound like.

ACTIVITY

Levels of formality

In English, you can express the same idea in a number of different ways, depending on how formal or informal you want to sound. For example, how would you ask where the toilet was in the following contexts?

> Asking your interviewer, at an interview
> Asking a stranger in a pub
> Asking your best friend in a disco or club

Different levels of formality in English are often related to foreign loan words: French- or Latin-based words are often more formal than Anglo-Saxon words. Here are some examples:

French/Latin origin	*Anglo-Saxon origin*
perspire	sweat
expire	die
desist	stop

Now read through the following passages, which are three accounts of the same incident. The passages represent three different levels of formality, or styles:

- A very informal style containing lots of slang and dialect (from the Manchester area)

- An average Standard English style

- A highly formal Standard English style

Read the passages aloud.

THE BARNEY

T'other day there were a barney at paper shop between Mr Arnold Higginbottom and Mrs Nora Grimshaw. Mr Higginbottom had been chunnerin' on about Mrs Grimshaw's two kids who'd been playing footy in the entry and gawping over his fence when they lost their ball.

Anyroad, Mrs Grimshaw were gobsmacked about this. She said, 'What ya' mitherin' about, ya' lemon? Why are ya' gettin' so nowty?' He said that he'd just legged it back from t'factory and he were feeling dead powfagged and he were pig sick of them moping about his yard. He said she should of leathered 'em in the first place and she'd better tell them if they did it again they'd get done.

She said that if it came to complaints, she had a few of her own. His moggy had been messing in her spud patch and had been scratting around there for weeks.

He told her to put a sock in it, and upped and went out of the shop. He barged into an old bid what were trying to come in, knocked his hat off his head and shoved him to the ground. The old codger started whingeing straight away and were carted off to the hospital where they gave him a full check-up. He were shook up and flapping, but apart from a sore lug'ole he were sorted, so they give him a brew and sent him off home for a kip.

continued

THE ARGUMENT

The other day, there was an argument at the newsagent's between Mr Arnold Higginbottom and Mrs Nora Grimshaw. Mr Higginbottom had been complaining about Mrs Grimshaw's two children who had been playing football in the alleyway and staring over his fence when they lost their ball.

Anyway, Mrs Grimshaw was surprised about this. She said, 'What are you complaining about, you silly man? Why are you getting so cross?' He said he had just hurried back from the factory and he was feeling really tired and he was fed up with them hanging around in his yard. He said she should have smacked them in the first place and she had better tell them that if they did it again, they would be punished by him.

She said that if it came to complaints, she had a few of her own. His cat had been digging in her potato patch and had been scratching around there for weeks.

He told her to be quiet and left the shop. He bumped into an old man who was trying to come in, knocked his hat off his head and pushed him to the ground. The old man started complaining at once, and was taken to the hospital where they gave him a full check-up. He was very shaken and flustered, but apart from a sore ear he was all right, so he was given a cup of tea and sent home for a nap.

THE ALTERCATION

Recently, there was a fracas at the newsagent's between Mr Arnold Higginbottom and Mrs Nora Grimshaw. Mr Higginbottom had been levelling accusations about Mrs Grimshaw's two offspring who had been engaged in certain team games in the communal passageway and looking fixedly over his fence when they mislaid their ball.

Notwithstanding, Mrs Grimshaw was absolutely incredulous at this. She retorted, 'What is the basis of your grievance, you asinine nincompoop? Why are you becoming so irate?' He responded by stating that he had just returned post-haste from the manufacturing establishment and he was feeling utterly fatigued and he was extremely disgruntled to find them loitering on his property. He said that she should have thoroughly reprimanded them at the outset and he would be obliged if she would communicate the fact that if they persisted in their activities they would be chastised by him.

Her riposte on the subject of grievances was that, in point of fact, she had a number of objections herself. His feline companion had been unearthing her King Edwards and had been disturbing the terrain for some time. He asked her to desist and made his departure from the premises. He collided with an elderly citizen who was attempting to enter, causing his millinery to be dislodged and jostling him to the ground. The old gentleman expostulated immediately and was accompanied to the infirmary where a thorough examination was conducted upon him. He was extremely tremulous and palpitating, but aside from an injured auditory organ he was in a satisfactory condition, so he was offered a hot beverage and was despatched homewards to rest and recuperate.

continued

What different impressions do you get of the person telling the story, from the differences in language used? How did it feel to read or listen to the accounts – for example, were the passages read in particular accents?

Try to pick out:

- dialect words and phrases from the first passage that help to give the account an informal, local flavour;

- words and phrases from the third passage that help to make the passage sound very formal (and comical in this case, because the incident would not normally be described in such a high-flown way).

ACTIVITY

In the jumble of words below there are 38 pairs which have the same or similar meanings (synonyms). Try to match up these pairs: each pair has an informal sounding word (from Anglo-Saxon/Viking languages) and a formal version (from French/Latin languages). Arrange the words in two columns under the headings:

Anglo-Saxon/Viking Words (informal)   *French/Latin Words* (formal)

snag	employment	smelly	endure
meet	offensive	casserole	attire
trousers	sweat	living room	know-how
friendly	worker	residence	job
odorous	make	clothes/togs	encounter
preserve	assignment	loving	keep
cuisine	date	house	stew
talk	employee	drink	underwear
amicable	occupation	give out	manufacture
perspire	expertise	lounge	kecks
melodious	converse	task	imbibe
commence	speed up	graveyard	launder
die	disseminate	work	booze
assistance	lingerie	impediment	cookery
meeting	rude	expire	amorous
accelerate	wash	cemetery	start
put up with	driver	alcohol	help
intoxicated	coiffure	tuneful	appointment
hair-do	rendezvous	drunk	chauffeur

(Answers are on page 71.)

When you have finished, use some of these words to write two accounts of the same incident: make one version very informal – include some dialect terms – and the other version very formal. Be prepared to read your different versions out to the whole group if you are working in a class situation.

Websites http://www.onelook.com

http://www.facstaff.bucknell.edu/rbeard/diction.html

Language level IV: Grammar This level relates to the structural patterns or 'rules' that any language has, in order to relate items to each other. If you imagine language as a wall, then the words and phrases are bricks, and the grammar is the cement that holds them together in certain patterns. Different walls – different languages – are patterned in slightly different ways.

Young children learn grammatical patterns along with vocabulary and sounds at a very early age; your own knowledge of grammar will be extensive in terms of your use of language, although you might not be able to analyse the structures you use in an abstract way.

There are many different aspects of grammar which could contribute to meaning in a text. Below are some of the aspects of grammar that it might be useful to consider as a starting point.

Verbs:
(a) *Tense* – Although writers may vary the tense of the verbs they use for reasons of style, certain types of writing tend to use particular tenses as their norm. For example, narratives often use the past tense (I came, I saw, I conquered), while generalised descriptions tend to use the present tense (The whale is the largest living mammal).

(b) *Modality* – Texts that concern themselves with ideas about possibility, probability or certainty – for example, weather forecasts, horoscopes, teachers' reports – often use modal verbs. These are verbs such as may, can, will, might and must. Modals are used with other, main verbs to express a range of meanings, including the likelihood of occurrences. For example, 'it will rain tomorrow', 'you may meet someone new', 'he might pass his exam'.

(c) *Aspect* – Different verb structures can express important ideas about the nature of the action. For example, the continuous aspect (I was writing) can suggest prolonged activity where the simple aspect (I wrote) can suggest brevity or a one-off action. The 'ing' form of the verb is called a participle.

(d) *Voice* – Some texts – particularly scientific reports – use passive verb structures rather than active ones. Where an active sentence specifies the doer of an action (the subject), in a passive sentence this element can be omitted. For example,

active: I conducted an experiment
passive: An experiment was conducted

Passive verbs are often associated for this reason with an impersonal effect.

Sentence types – Different sentence types fulfil different functions. For example, a set of instructions will rely on command sentences (sometimes called *imperatives*), whose function is to give orders; a report may use many statement sentences (sometimes called *declaratives*) to convey facts; a piece of persuasive writing may ask the reader questions, to get him or her thinking (*interrogative* sentences) or use exclamations (*exclamatory* sentences) to express certain feelings and attitudes.

Word order – In English, certain types of words have to precede others. For example, single adjectives usually have to go in front of nouns (there are exceptions where phrases have been borrowed from other languages, e.g. 'court martial', from French). Word order can also be manipulated, to a certain extent, to emphasise particular elements. For example, if writers want an item in a sentence to stand out, they will sometimes try to put it at the end, which is a point that we notice particularly – this is called the *focus* position.

Modification – This refers to the way that words and phrases can be added to others to build up descriptive information, for example, adding adjectives and adjective phrases to give further information about nouns (e.g. a tall, Victorian building with rotting gutters).

When items go before the noun, this is called *pre-modification*. When they go after it, this is called *post-modification*.

Omission of items – Certain types of writing may involve the omission of some parts of sentences. For example, newspaper headlines often omit the articles 'a' and 'the'.

Pronouns – Pronouns (I, you, he, she, it, we, they, one) are among our smallest words but can be used to achieve powerful effects in texts. For example, whether I write as 'I' or 'we', whether I address you directly as 'you', whether I use 'he' or 'she' in the examples of sentences I include, all make a considerable difference to how this text I am writing is perceived.

Grammatical ambiguity – Ambiguity in texts can arise because the reader/listener can see more than one way to understand the grammatical patterns. This in turn can be because grammatical elements can shift their word class membership. For example:

nouns can become verbs: a parent/to parent a child
verbs can become nouns: to eat/let's have some eats
adjectives can become nouns: comic/a comic
verbs can become adjectives: I must see that film/it's a must-see film.

So one phrase can sometimes be read in more than one way because certain of the elements could be in more than one word class. For example, in:

visiting relatives can be boring

'visiting' can be an adjective describing 'relatives', or a verb meaning 'to visit relatives', where the subject is not specified.

Another way grammatical ambiguity can arise is through uncertainty about where the boundaries are in a phrase. For example, in:

slow children crossing

does 'slow' describe the children, or is there a boundary after the word 'slow', leaving 'children' and 'crossing' to go together?

ACTIVITY

Look at the data below, and try to answer the questions attached.

NEWSPAPERS
(a)   What makes these headlines ambiguous?

GIANT WAVES DOWN TUNNEL
(article about the flooding of the cross-channel tunnel)
OLD TRAFFORD NURSE ATTACKED
GENERAL FLIES BACK TO FRONT
NINE YEARS FOR DROWNING BABY

(b)   What technique is being used in these newspaper articles?

the ex-world welterweight champion boxer Lloyd Honeyghan
Mrs Mills, a 25-year-old mother of three who is now separated
that old Labour levelling tendency
Mick Hucknall, celebrity playboy and purveyor of white-boy soul to the suburbs

ADVERTISEMENTS
What techniques are being used in the following?

(Boddington's Beer) If You Don't Get Boddies, You'll Just Get Bitter
(Kenco Coffee) Everything We Know About Coffee, in an Instant
(Men's face moisturiser – picture of a man's face looking to camera)
His good looks may be irritating, but his skin isn't

LITERATURE
This extract is the opening page of a children's book called *The Piggy Book*, by Anthony Browne.

The book is about sexism and features a family where the mother is very downtrodden. How is that message put across by the grammar of this extract, as well as the vocabulary?

'Mr Piggott lived with his two sons, Simon and Patrick,
in a nice house, with a nice garden,
and a nice car in the nice garage.
Inside the house was his wife.'

continued

What grammatical features are noticeable in these different types of instructions? (Also think about the other language levels you have studied so far, and comment on any language features that relate to them – for example, graphology, semantics.)

Computer Manual
If your computer system does cause interference to radio or television reception, try one or more of the following measures:

- Turn the TV or radio antenna until the interference stops
- Move the computer to one side or other of the TV or radio
- Move the computer further away from the TV or radio
- Plug the computer into an outlet that is on a different circuit from the TV or radio

Knitting Pattern
Cast on 53 [53 : 58] sts with No. 10 needles. Work 11 rows rib as for lower edge.

Next Row: P1 [1 : 5], *page 2, inc in next st, rep from *to last 1 [1 : 5] sts, p to end 70 [70 : 74] sts. Change to No. 8 needles. Work 6 rows, having 3 [3 : 5] sts each end of needle in d.m.st, and central 64 sts in patt. as Back from *to**. Working extra sts in d.m.st, inc at both ends of next and every 6th row following until there are 96 [102 : 106] sts. Proceed to 19 inches, ending after a wrong side row.

Recipe (Holiday Confetti Bread)

1 cup milk	$^1/_2$ cup sugar
1 $^1/_2$ tsps salt	6 tblsps butter
1 tsp lemon peel	$^1/_2$ tsp ginger
1 package active dry yeast	$^1/_4$ cup lukewarm water
2 eggs, well beaten	4 cups sifted all-purpose flour
1 $^1/_2$ cups mixed candied fruits, cut	$^1/_2$ tsp allspice
into small pieces	$^1/_2$ cup slivered almonds
2 tblsps flour	

Scald milk; add sugar, salt, butter and spices. Cool to lukewarm. Dissolve yeast in water. Stir in beaten eggs and milk mixture. Add the 4 cups flour and stir until moistened. Cover and set in warm place and let rise until double in bulk, about 1 1/2 hours. Dredge fruit with the 2 tablespoons flour. Add fruit and almonds to batter; beat 2 minutes. Push into greased 2-quart mould. Set in warm place and let rise until double in bulk, about 1 hour. Bake in 350 F oven 1 hour. Makes 1 loaf.

continued

ENGLISH AS A FOREIGN LANGUAGE
Look again at the penfriend's letter on page 20. What aspects of English grammar is the writer having problems with?

If you have been working in small groups, share your findings with the whole group. If you have been working individually, make some notes on your findings for your file.

Language level V: Discourse

When we encounter a written text, the features of language studied so far act as clues to help us answer some important questions about the text as a whole. For example, as we are reading, we may ask ourselves certain questions:

- What type of text is this? For example, is it an advert, a story, an information text?

- Where has it come from? For example, a private individual or a large company?

- What is it for? For example, to persuade, to inform, to entertain?

- Who is communicating? For example, what kind of person appears to be addressing me (i.e. the 'narrator') and why is he or she speaking to me in a certain way?

- Who is being addressed? (i.e. the 'narratee'). For example, does the text assume I am a certain kind of person? Do I feel included or excluded by the way the text communicates?

These larger questions are all concerned with what is called the 'discourse' level of texts – that is, the way a text works as a whole piece of communication. These workings can be very complex, so don't assume that the questions above can lead to simple answers. For example, it's a well-known feature of adverts that they can 'impersonate' other texts. So you might encounter an advert that is constructed as a story. Texts can have many purposes at once: they can try to persuade through entertainment, for example. Sometimes, texts address us via a number of different voices – different narrators constructed to give us, as readers, particular and different viewpoints. And it can be the case that texts don't simply include us or exclude us. For example, an advert might be pitched at an idea of myself that I want to aspire to, or that I fear I might really be – in other words, an abstract 'construction' of me. In this case, the text is not ignoring me, but nor is it aiming itself directly at me: it is aiming at a point slightly away from me, but one that it wants me to reach.

If, as readers of texts, we are coming to such conclusions as the above, then as analysers of texts we need to probe the level of discourse as much as possible. Otherwise, we will be left with a checklist of features of language, but never add up these features to form any satisfying picture of how the text really works in action.

ACTIVITY

Read the extracts that follow, with all the different language levels in mind:

Graphology; Phonology; Semantics; Grammar; Discourse.

Write down some notes on what is distinctive, or worthy of comment, about the extracts from the point of view of each language level. Think particularly carefully about the level of discourse, and come to some conclusions about how the whole texts work as pieces of communication. Remember that your notes on the other levels of language are your evidence for the bigger discourse questions you will be asking and answering.

(Notes on the first two extracts are given on pages 71–73.)

9.25 This House Possessed

This movie spooky has a plot corny: sensitive nurse Sheila gets bad vibes in the mountain mansion where she's helping rock star Gary to get over his breakdown, and a series of sinister incidents convinces her that the place is alive. Lisa Eilbacher and Parker Stevenson are the cut-off couple, with Slim Pickens as the singer's manager. Hollywood veteran Joan Bennett as the old dear who knows what it's all about. Made in 1981.

The Guardian TV listings

INVENTORY
        you left me
            nothing but nail
            parings orange peel
            empty nutshells half filled
            ashtrays dirty
            cups with dregs of
            nightcaps an odd hair
            or two of yours on my
            comb gap toothed
            bookshelves and a
            you shaped
            depression in my pillow.
                    Liz Lochhead

continued

I tried to kill myself.
My Mother said I should have tried harder.

This is Alexandra's story.

'I don't know who hated me more, my mother or myself.

One of my first memories was waking up from a bad dream. I was so upset I had to disturb her dinner party.

Her face lit up when she saw me. She kissed me. I felt so happy. Then she took me up to bed and shut the door.

Her face transformed. She told me that having me was the worst thing she ever did. I was so confused.

5 years later when my father left us, I tried to put my arms around her but she pushed me away saying he'd left us because he couldn't stand the sight of me. I decided to kill myself.

When I woke up in hospital my mother was there. "You should have tried harder" she said. When I told the nurse, she was horrified. But I'm coming to terms with it all, thanks to the counselling she suggested.'

As Alexandra's story shows, child abuse isn't just about sexual assault and physical brutality.

For example, being constantly criticised, shouted at or ignored may be less obvious kinds of abuse. But they can make a child grow up feeling worthless, depressed, even suicidal.

Because of all this, the NSPCC is launching a campaign called 'A Cry For Children.' It's a cry to everyone to

stop and think about the way they behave towards children.

To recognise the impact that any form of cruelty can have on a child. And to realise that the way children are treated affects their whole lives.

Please answer the cry.

If you, or someone you know is suffering abuse, call the NSPCC Child Protection Helpline on 0800 800 500.

Or if, after reading this, you would find more information helpful, please call us on 0171 825 2775.

NSPCC
A cry for children.

Genres of writing

'Genre' is a word of French origin, meaning 'type' or 'category'.

As a language user, you already know a great deal about different types or categories of writing. The work you have done in the previous section on different levels of language will help you in talking explicitly about written genres in the activities that follow.

ACTIVITY

Genre game

Below are some extracts from different written genres.

Decide which genre of writing each of the extracts has been taken from. Then use your own knowledge to make up ten more.

If you run out of ideas, think about some of the extracts you studied in the activities on language levels.

When you have finished your work in small groups, each group should circulate its sheet round the other groups, for them to guess.

If you have been working individually, test out your data on some informants, to see how far others are able to guess the genres.

EXTRACTS

1. In 1516, the Treaty was finally ratified by all parties ...

2. A substance was placed over a bunsen burner and heated ...

3. With Many Happy Returns ...

4. In Loving Memory of ...

5. She came to him in an agony of passion and tenderness ...

When you have finished, consider the following:

• What were some of the clues you used to identify the extracts? Did these clues correspond with the language levels you studied previously?

• Which, if any, of the extracts were hard to guess, and why?

ACTIVITY

Dimensions of written variation

Now, using the headings below, fill in some details for the genres of writing you covered in the previous exercise.

Sometimes, the audience could be a number of different groups. In that case, it may be useful to fill in a range of possibilities.

Audience   *Purpose*   *Format*

If you have been working in groups, gather all your results together at the end on one large A3 sheet, and pin it up for everyone to see.

If you have been working individually, think carefully about the range of writing you have accounted for. As well as making you more aware of the range of written genres for purposes of investigation, your collation will also help you with the planning of any pieces of your own writing.

Decoding spoken texts

Just as any reader knows a great deal about written texts and how they work, any speaker knows about spoken texts – their rules, as well as their features.

As speakers and listeners, we are not aware of all the different components which make up spoken communication; for example, in practice, we do not divorce intonation from vocabulary, or grammar from the physical gestures that we use.

However, for purposes of analysis, we need to break speech down into manageable elements, in order to see what part each of these elements plays in the whole act of communication. For this reason, this section is subdivided as follows:

1. It's the way that you say it
2. The spoken alphabet
3. The speech context
4. The grammar of speech
5. Spoken vocabularies
6. The rules of interaction

Each subsection provides activities and speech extracts which illustrate the element being analysed; towards the end, you will have the chance to put these elements together, to see how they all work at once.

1 It's the way that you say it

Obviously, there is no such thing as punctuation in speech.

Written punctuation marks and other graphological devices are an attempt to suggest, for writing, what our voices do when we are talking.

ACTIVITY

> Read through the following examples and, paying particular attention to the punctuation in each, decide how the examples should be spoken when read aloud:
>
> (a) Hello. I said HELLO!
>
> (b) What on earth do you think you are doing?
>
> (c) She rose from her chair, walked to the door, and slammed it behind her.
>
> (d) No!!
>
> (e) It's pronounced 'Hospital'.
>
> (f) Well, I don't know, really ...
>
> (g) I didn't *mean* to do it – it just *happened*.
>
> (h) You can receive benefit (or the equivalent rebate) if you claim before the stated period.
>
> (i) He entered the room. The atmosphere was tense. Nobody spoke.
>
> (j) Take the following: a tablespoon of grated Parmesan; three eggs, beaten lightly; a generous handful of basil, chopped coarsely; and a pinch of salt.

If you were able to read the signals given by the punctuation marks and to turn them into vocal effects as you spoke, then you know a lot about the relationship between *written symbols* and *spoken intonation*.

When you want to transcribe a piece of real speech and represent in some detail how it actually sounded, however, written punctuation is a rather crude convention.

Real speech is not neat and tidy, like writing; it does not necessarily fit the written conventions of sentences (or even words, since we speak in bursts of sound); and written punctuation cannot do justice to the subtleties of intonation, pauses, and so on.

Transcription conventions

Transcription conventions vary, but in many transcripts, boundary lines / are used to show pauses between utterances. Sometimes, transcribers also indicate longer pauses than average by using brackets, with a dot to represent a brief pause (but one which is longer than the 'norm' for a particular speaker) (.) and numbers of seconds for even longer pauses (1), (2).

Pauses can be *filled* as well as silent, with speakers making noises such as 'er' or 'um' to keep their speech continuous.

ACTIVITY

> Read through the child's speech below, which has been marked with boundary lines. Read it aloud to hear how it might have sounded. Then decide:
>
> • how the spoken divisions differ from the conventional sentence boundaries that you would have in writing
>
> • what the function is of the filled pauses.
>
> er my name is Keir/and I'm six years old/and my best friend/is Matthew Harris/I go to Woodcote um/Woodcote Primary School/I think it's horrible/we're always doing work/my brother's um/three years old/and he's naughty/my worst enemy/is Darren Hunt/my second worst enemy/is David Harding/my third worst enemy/is Kalvin Racy

Intonation

Intonation is a melody that we all sing.

If you compare speech with a song, then intonation is the backing tune to the words.

Intonation is our 'oral punctuation'; our voices signal many meanings, including the message that one part of an utterance is ending and another is about to start. This ensures that the ambiguity which is often present on the page, in written material, is avoided in speech: for example, the sentence 'The terrorists wrecked the factory by blowing up the pipes' would never be ambiguous when spoken.

The divisions we make with intonation are called *tone groups*. These are the sections that you were looking at and reading aloud in the previous activity. Each tone group has one main stress, or intonation point. This is marked on the main stressed syllable in the group, which is called the *nuclear syllable*. Sometimes there is also a lighter stress before the main one, which is where the whole tune actually begins. This is called the *head*.

ACTIVITY

> Read through the utterances below, and mark:
>
> • where the main divisions (tone groups) would have been in each utterance
>
> • where the nuclear stress would have occurred for each tone group, with a * by the syllable
>
> • if there is a secondary, earlier stress before the main one (the head), with a ´.
>
> The first two examples have been completed, to start you off.
>
> *continued*

N.B. Assume that each utterance is conveying a straightforward, rather than any unusual, meaning. (Answers are on page 74.)

(a)   She 'ran to the *station/and 'caught the *train

(b)   He 'fried the *onions/and 'chopped the to*matoes

(c)   She went to London but wished she hadn't

(d)   They came in early and worked until lunchtime

(e)   He hurried along to the centre of town

(f)   I got a puncture while I was driving

(g)   After they left I was exhausted

(h)   She went to the bank at the end of the road

(i)   He hailed the bus and jumped on

(j)   Before I go I must make a phone call

Here is the transcript you worked on previously, with nuclear stress and heads marked. Try reading this aloud, interpreting the marked stresses and pauses:

er my 'name is *Keir/and I'm 'six years *old/and my 'best *friend/is 'Matthew *Harris/I 'go to *Woodcote um/ 'Woodcote *Primary School/I 'think it's *horrible/we're 'always doing *work/ 'my *brother's um/ 'three years *old/ and 'he's *naughty/my 'worst *enemy/is 'Darren *Hunt/ 'my *second worst enemy/is 'David *Harding/ 'my *third worst enemy/is 'Kalvin *Racy

Common tone groups

There are certain structures that often occur in tone groups, and there are examples of these in the data you have been studying.

Subject – Verb: My best friend/is Matthew Harris; My worst enemy/is Darren Hunt

Co-ordination (linking with 'and'): They came in early/and worked until lunchtime; My brother's three years old/and he's naughty

Subordination (linking with 'while', 'after', 'when', etc.): I got a puncture/while I was driving; After they left/I was exhausted

Adverbials (words or phrases giving information about where, when or how actions were done): He hurried along/to the centre of town; She went to the bank/at the end of the road

Interjections and formulaic expressions: Words and phrases like 'Well!' 'Now then!' Oh no!' 'Good morning!' often have their own tone groups. Sometimes this applies to fillers such as 'um' and 'er' as well.

Intonation tunes

As well as placing a much heavier stress on the *nuclear syllable* in any tone group, we also sing a particular tune.

ACTIVITY

To get some ideas of the tunes we sing, have a conversation which involves no words, but try to get your meanings across to another person by using intonation alone.

Instead of speaking in words, use numbers – any will do, but two or three figure numbers will give you more of a chance to make your tune clear.

Here are some of the meanings that you can try to convey to others:

anger	disagreement	sadness
issuing a command	sympathy	reassurance
joy	sexiness	being impressed
asking a question		

When you have finished your conversation, consider the following:

- How much meaning did you feel you could convey, using intonation alone?

ACTIVITY

We learn how to produce intonation tunes very early in life – when we are small babies. But there is a difference between being able to use intonation and being able to analyse it. The information that follows is all about the different intonation tunes that we use. Because the area is a complex one, and because it is being presented here on paper, you may well need to go over it several times. It is particularly important that you try out the various sounds as you go along.

There are seven main types of intonation *tune* in English, and each is marked in an individual way. There are also different types of *head*.

1. HIGH FALL: `Ann

This is where the voice starts high in pitch, and falls downwards.

It is very commonly used for straight statements, questions, and for commands, and, in these cases, takes a high head:

     I ´really `like her
     I'm ´going on `Tuesday
     ´What are you `doing?
     ´How much is `that?
     ´Go a `way!
     ´Open the `door!

2. LOW FALL: ˌAnn

This is where the voice starts quite low in pitch, and falls downwards.

It is less commonly used that the high fall tune, but can be used to suggest the same sorts of meaning. However, low fall often conveys less enthusiasm and commitment because it sounds less energetic.

continued

Therefore it can mark weariness and irritation:

> What are you ˎdoing?
> Go a ˎway!

3. HIGH RISE: ˊAnn

This is where the voice starts high, and rises.

It can suggest shock and disbelief:

> What did you ˊsay?
> ˊWhat?

It is also the tune that people often employ when they are calling up the stairs to someone who should have been downstairs long ago:

> ˊJohn (Get up this minute, you're breakfast is getting cold!)

4. LOW RISE: ˏAnn

This is where the voice starts low, and rises.

It can create a caring and reassuring meaning, and, as such, is often used when adults speak to children (or animals!) in statement or question form. In these cases, it starts with a high head, so the beginning of the utterance is kept high:

> There's ˊno need to ˏworry
> ˊDid you have a nice ˏdinner?

When it is used with a low head, it can sound very grumpy:

> ˏYou can't do ˏthat

5. MID LEVEL: >Ann

This is the sound that we make when we want to imitate computerised or synthesised voices. It is right in the middle of our voice range, and to speak in it for any length of time makes us sound very robotic. It is the tune that teachers use when they are calling the register in a monotone:

> >Robert
>
> >Lisa
>
> >Janet
>
> >Peter

6. RISE-FALL: ^Ann

This is where the voice starts low, rises high, then falls again markedly in a short space of time. It is the sound we make when we

continued

are conveying a 'gossipy', 'well, you don't say', 'you are a naughty girl', 'ooh, I'm impressed' meaning:

> Get ^her/she got a dis^tinction

7. FALL-RISE: ᵛAnn

This is the opposite of the above. The voice starts high, falls, then rises again in a short space of time.

The tune is said very frequently in English for a range of meanings. It is notoriously difficult for foreign learners to acquire, as it is not a straight movement of the voice, but a 'wobbly' sounding tune. It is most clearly expressed as a tune alone, with no accompanying words, by children when they are doing what we call 'whining'. Imagine a young child that you know has been told that he or she can't go to the park after all, because it's raining. The child utters a long-drawn out 'oh' sound with a whining intonation. This is an exaggerated fall-rise tune.

While in children's speech it often characterises disappointment, in adult dialogue it frequently expresses contrast, contradiction or warning, and, in these cases, is sometimes used with a falling head (↘):

> I ↘like his ᵛwife (= but I don't like *him*)
>
> (He's thirty, isn't he?) ↘Thirty ᵛfive
>
> (Dave couldn't make it) But ᵛPaul could
>
> You'll be ᵛlate

It can also be used for urging people on:

> Please ᵛtry

ACTIVITY

Consider the following:

- What is the difference between these three ways of saying good morning?

  ´Good `morning
  ´Good ˎmorning
  ´Good >morning

- Lists have a particular sequence of intonation tunes: all items until the last have a rising tune, then the final item on the list has a falling tune, to signal completion:

  There's ˎwhisky/ ˎgin/ ˎbrandy/and `wine.

Now mark these slightly more complicated lists – from Tony Blair's pre-election press conference to launch Labour's 'Education Action Zones' (15/4/97). Listing (often in threes, but also in contrasting pairs) is a persuasive device often used by public speakers. It is important that the speaker gets the intonation sequence right, in order to sound convincing. More practically, at rallies such as party conferences, uncertain or wrongly placed intonation can confuse the audience about when to clap.

(i)   Under the Tories a brain drain, with Labour the brains coming home.

(ii) The dividing lines are clear on education in this campaign – positive versus negative, future versus past, many versus few.

- What is the difference in meaning between the falling and rising tunes in the following tag questions?

  ´Nice `day/ `isn't it
  ´That's `right/ ˎisn't it

- What would be the difference to the above if you reversed the intonation patterns, using rising tunes in the first, and falling in the second?

The following conversation has the nuclear syllables marked in the wrong places, and the tunes marked are not necessarily the correct ones.

Read it aloud as it is written, to get a sense of why it is wrong.

Then write it out again, putting stresses in the right places, with the right tunes attached. The conversation is between two teachers, at exam time.

(Answers are on page 74.)

A: `what do I do/with these exam `papers

B: `sort them into bundles/ `take them to the classrooms/and `dish them out

A: ^OK/ `what else

B: `that's about it/ `I think/how about `using your own initiative

A: `that's your job/I'm just a dogs `body

Speed and volume

Musical tunes are not just about the notes being played. Factors such as the speed of the music, and the relative loudness of various parts of it, also play a part in how we interpret the melodies we hear.

The same is true of our voices. We vary the speed and volume of our speech to signal certain types of meaning, or as part of particular strategies in dialogue.

Transcribers sometimes use musical notation to describe speed and volume, as follows:

Speed: allegro (fast)
        lente (slow)
        accelerando (accelerating)
        rallentando (slowing down)

Volume: forte (loud)
        piano (soft)
        crescendo (becoming louder)
        diminuendo (becoming softer)

ACTIVITY

Read through the dialogue below, which is the opening part of a telephone conversation between two friends and work colleagues. A has just bought a computer and is phoning B in order to get some help with a problem. The phone call follows two previous messages which A had left on B's answerphone machine, asking for help on the same topic.

To start with (line 1), A begins to leave a message on B's answerphone, believing B to be out. Then B cuts into the message by picking up the handset (line 2).

When you have read through the dialogue, think about the following:

• How do the different levels of volume and speed contribute to the meanings being conveyed by the speakers?

• A varies considerably more than B, both in volume and speed. These variations are nothing to do with any absolute differences between the people as speakers in general: the variations demonstrate aspects of how the speakers are managing this particular conversation. How might these variations be explained, in this specific context?

• Can you see verbal aspects of the dialogue that work alongside the acoustic features you have been studying?

continued

* = overlapping speech

Line No.
1   A: Sam it's Chris

2   B: (piano) hello

3   A: (crescendo) oh hi

4   B: (piano) I'm here

5   A: (forte) thank god you're here

6   B: (crescendo) why what are you doing

A: (first part of utterance is inaudible because of laughter and very loud speech volume) I'm on this computer I've spent all evening on it I've got icons all over the place they're coming out of the woodwork

B: (lente) I can't hear you very well what are you saying

A: (rallentando) I'm doing something that is every time I open a *file*

B: (lente) *yeah*

A: (rallentando) and all I'm trying to do is just explore and have a look at um all these wonderful different things just have a look in and I can close it again but (accelerando) what's happening is every time I go under file and look at it it's putting it in my letters thing so I've got all these icons in my letters folder

B: that's 'cos you're saving it in letters

A: pardon

B: (lente) that's 'cos you're saving it in letters

A: (accelerando, crescendo) but I don't want to do that and what's awful is every time I try to get rid of them it says it's going to put them in the recycle bin

B: (lente) yeah well it will that's all right

2 The spoken alphabet

You have seen in the previous section that written punctuation cannot represent the many subtle meanings conveyed by intonation.

In the same way, the written alphabet cannot represent the many different ways in which one word could be pronounced by people from different parts of the country – in other words, *accent features*.

In the real speech context, speakers are very skilled at registering and interpreting accent variations – both the way in which speakers from different areas differ from each other, and the way in which one speaker can vary his or her accent during the course of a conversation, to signal certain meanings.

Because writing is a standardised medium (i.e. everyone learns and uses the same system), it does not easily represent such subtle differences.

The *phonemic alphabet* is a way of transcribing the actual sounds that people make. A *phoneme* is 'a single, distinctive sound'. The phonemic alphabet is better than the spelling alphabet for describing sounds because one spelling can have different sounds – <u>cough</u>, thr<u>ough</u> – and one sound different spellings – m<u>ea</u>t, f<u>ee</u>t.

The phonemic alphabet on page 44 is based on the RP – Received Pronunciation – accent (the accent often used by national newsreaders).

ACTIVITY

One person should read out the words from the list below and the other member of the pair should write them down as they were actually said. If you want to share the speaking and transcribing tasks, split the list in half, and each read 14 words.

If you are working individually, ask someone to read the words aloud to you.

The person reading the words should try to reproduce them in as natural a way as possible, and not try to use a more formal pronunciation than is usual. He or she may have to say the words several times in order for the transcriber to keep up. The transcriber needs to forget about the spelling of the words, and listen very carefully to the sounds that are actually being made.

WORD LIST

THERE	CITY	PUTT	THREE
BUS	AFTER	PULL	OFF
CAR	FUR	POOL	BOTTLE
BATH	FAIR	POOR	HOSPITAL
DOWN	WHO	FAR	ROAD
COT	HUGH	FIRE	ABOUT
DANCE	PUT	SINGING	THEM

THE ENGLISH PHONEMIC ALPHABET (Based on the RP accent)

LIST OF SYMBOLS

English vowels

1. i   as in s<u>ee</u>
2. ɪ   as in s<u>i</u>t
3. e   as in s<u>e</u>t
4. æ   as in s<u>a</u>t
5. ɑ   as in c<u>a</u>lm
6. ɒ   as in n<u>o</u>t
7. ɔ   as in b<u>ou</u>ght
8. ʋ   as in p<u>u</u>t
9. u   as in b<u>oo</u>t
10. ʌ   as in c<u>u</u>p*
11. ɜ   as in b<u>i</u>rd
12. ə   as in <u>a</u>bout
13. eɪ as in pl<u>ay</u>
14. əʋ as in g<u>o</u>
15. ɑɪ as in m<u>y</u>
16. ɑʋ as in n<u>ow</u>
17. ɔɪ as in c<u>oi</u>l
18. ɪə as in h<u>ere</u>
19. ɛə as in th<u>ere</u>
20. ʋə as in cr<u>ue</u>l

English consonants

1. p   as in <u>p</u>ut
2. b   as in <u>b</u>ut
3. t   as in <u>t</u>en
4. d   as in <u>d</u>en
5. k   as in <u>c</u>ome
6. g   as in <u>g</u>o
7. tʃ as in <u>ch</u>urch
8. dʒ as in ju<u>dg</u>e
9. m   as in <u>m</u>ake
10. n   as in <u>n</u>et
11. ŋ   as in lo<u>ng</u>
12. l   as in <u>l</u>ong
13. f   as in <u>f</u>ull
14. v   as in <u>v</u>ery
15. ɵ   as in <u>th</u>in
16. ð   as in <u>th</u>en
17. s   as in <u>s</u>at
18. z   as in <u>z</u>eal
19. ʃ   as in <u>sh</u>ip
20. ʒ   as in mea<u>s</u>ure
21. r   as in <u>r</u>un
22. h   as in <u>h</u>at
23. w   as in <u>w</u>ent
24. j   as in <u>y</u>et

*the RP pronunciation of 'cup'. Many northern speakers do not have this phoneme in their speech.

There is one more sound which is often used, but is not listed above because it is not a full sound. This is the *glottal stop*, and is written as ?. It is similar to the 'h' sound, but differs from it by being produced by closing the vocal cords, as you would when you pick up a heavy weight. It sometimes replaces other sounds in certain accents, for example,

Cockney speakers would say bɒ?əl
        compared with RP bɒtəl

ACTIVITY

Now compare your results with the RP accented versions of the words, given below:

- Which major differences in sound characterise the speech of your region?

- Are you aware of different ways of pronouncing some of the words on the word list, even though they may not be the pronunciations you yourself use?

WORD LIST

RP pronunciation		*Notes on other accents*
THERE	ðɛə	Liverpool = ðɜ; Patois = dɛə
BUS	bʌs	Northern = bʊs
CAR	kɑ	West Country = kɑr
BATH	bɑθ	Northern = bæθ
COT	kɒt	Scottish = kɔt
DANCE	dɑns	Northern = dæns
AFTER	ɑftə	Salford (Greater Manchester) = æftɒ
FUR	fɜ	Liverpool – both = fɜ
FAIR	fɛə	
WHO	hu	Norfolk – both = hu
HUGH	hju	
PUT	pʊt	Northern – both = pʊt
PUTT	pʌt	
POOR	pɔ	Scottish = pʊə
SINGING	sɪŋɪŋ	Many regional accents = sɪŋɪn
THREE	θri	Cockney = fri; Patwa = tri
BOTTLE	bɒtəl	Cockney = bɒʔəl
HOSPITAL	hɒspɪtəl	Cockney = ɒspɪʔəl
ROAD	rəʊd	Lancs/Yorks = rɔd
ABOUT	əbaʊt	Cockney = əbat
THEM	ðem	Cockney = vem; Patwa = dem

ACTIVITY

The work that you have done on sounds so far related to regional accent features, and has involved speakers reading items on a word list aloud. When people read isolated words aloud, they are usually very careful about their pronunciation, and the way they speak will differ from how they would be in informal conversation.

There are certain aspects of spoken language which come into play when we say words in a piece of connected speech, and these aspects are present whatever accent we happen to have. This activity will help you to understand these aspects of connected speech.

One person in each group should read the passage below onto a tape. The reader should try to read in as relaxed a manner as possible.

Then each group should transcribe the sounds made by the reader, using the phonemic alphabet. Work together on this, helping each other to decide exactly what sounds were made.

If you are working individually, ask someone to read the passage for you onto a tape.

PASSAGE

She found her handbag and pulled out a white handkerchief. Her secretary, Peter Andy, had made her a celebration meal: a platter of cold meats, hot-pot, followed by tinned peaches and cream, and about ten cups of coffee to sober them up after the champagne.

Her handbag gaped too wide for me to ignore the object which met my open, two-eyed gaze: a box from India or China, with delicate carvings and the initials 'E.E.N.'.

When you have transcribed the passage, look particularly at the following aspects of connected speech:

Assimilation

This is where a sound changes under the influence of another.

For example, how was 'ten cups' pronounced by the reader? Was it the following:

$$ten \ kʌps \ ?$$

If so, this is assimilation, as n becomes ŋ so that the mouth is in the right position to say the k in 'cups'.

Make a list of any other examples of assimilation you can find.

continued

You might check how the following were said. If they were said as the phonemic transcription below suggests, this means that assimilation was happening.

handbag	handkerchief	tinned peaches
hæmbæg	hæŋkətʃif	tɪmpitʃɪz

Elision

This is where a sound is left out from a word, when it would normally be pronounced if the word was said in isolation.

For example, how was 'secretary' pronounced by the reader? Was it the following:

sekrətri

If so, this is elision, as when the word is pronounced in isolation, people will normally say:

sekrətəri

Make a list of any other examples of elision you can find.

You might check how the following were said, paying attention to whether the sounds in bold type were actually pronounced:

col**d** meats   ho**t**-pot   **h**andbag   **h**andkerchief   **h**er   platt**er**

Liaison

This is where linking sounds are inserted by speakers, so that there is a smooth transition from one word to another. The sounds that are inserted are often /r/, /j/ or /w/.

For example, did the following have the linking sounds described below (in brackets)?

my open	India or China
maɪ (j) əʊpən	ɪndiə (r) ɔtʃaɪnə

If so, these are examples of liaison.

What about the two phrases in the passage, 'too wide' and 'two-eyed'? Is there a linking /w/ when two-eyed is pronounced? If so, how is it then distinguished from 'too wide'? Or is there no difference between them – is it just the context that enables us to tell them apart?

continued

Juncture

Where liaison is all about smooth transition and linking, juncture is about keeping sounds apart. You can see an example of why we might want to do this if you look at the name 'Peter Andy'. If you run the sounds together in this name, then you get something unfortunate:

$$/\text{pitərændi}/$$

If you were using juncture in this example, and keeping the two names apart, then you were probably closing your vocal cords as if you were about to produce a glottal stop, just before you said the name 'Andy'. Also, you will have been careful not to pronounce the /r/ at the end of 'Peter'.

Make a list of any more examples of juncture you can find. You might check how your reader said 'E.E.N.'.

Check your transcriptions against the notes above. (There may well be variations according to the different readings you were transcribing.)

Has this activity helped you to understand how connected speech can differ from words pronounced in isolation?

The aspects of connected speech you have been studying are present in the spoken language of everyone, whatever their accent may be, and become more noticeable as speakers relax and sound less formal.

Were the readers of the passage conscious of trying to be careful in their speech?

How could you collect speech that was less self-conscious than the reading of the passage?

3 The speech context

Speech differs from writing in being supported by the physical setting in which it takes place. Physical gestures, for example, often have a meaning in their own right (think of the behaviour of drivers in traffic jams, or the many gestures that are used in sign language by deaf speakers). Many physical gestures are also language specific: for example, a 'thumbs up' sign, which is innocent and positive for English speakers, is very rude in Spanish.

The physical setting has an effect on the language that is used in spoken communication. For example, the fact that speakers can see objects that are being referred to means that the full names of the objects don't have to be used. Words like 'these, those, this, that' are often all that are needed: these are called *deictics*, which means 'pointing words'. The fact that speakers register each other's meanings and feelings as speech takes place means that language doesn't always need to be fully explicit: much can be left out. For example, a speaker recalling an upsetting experience might say '... and then I felt, you know ...' The listener would fill in the details not fully explained. This missing out of details that are understood is called *ellipsis*.

ACTIVITY

Look at the extract below. What activity do you think the speakers are engaged in?

(Asterisks denote overlapping speech.)

A: if you start here/um/what you can do/is take this up gradually/stopping to just check/this is even/*then*

B: *what* about this side/how do you keep it/out of the way

A: well/just get one of these/and clip it up/and forget about it

B: right/so take that up *gradually*

A: *yeah*/and make sure it's level/don't get carried away/you can always come back to it later

B: so then which way do you go

A: up here/then the same the other side/the top and front last/then you can see for yourself/how the shape's coming on/and the top and front/should take their shape/from the rest/that's the skill really/judging as you go/even though you have the main idea/in your head.

(Answer on page 74.)

Consider the following:

- What clues did you get about the activity the speakers were engaged in, to help you to understand all the references to 'here', 'there', 'this' and so on?

ACTIVITY

Write a short conversation between two or more speakers who are engaged in a particular activity.

Imagine that the speakers are talking as they are working, referring to what they are doing. Try not to give the answer away too easily: use examples of deictics and ellipsis.

When you have finished, read your conversation aloud to the other groups, for them to guess what your speakers were doing. If the other groups cannot guess, act out your conversation with the appropriate physical gestures.

If you are working individually, read your conversation aloud to someone, and ask him or her to guess the activity.

ACTIVITY

Record some speakers who are engaged in a particular activity – for example, cooking a meal, playing a game, constructing something. Transcribe the results. How does your real data compare with the fictive version you made up previously?

4 The grammar of speech

Speech is just as structured as writing, but its 'rules' are sometimes different.

As speakers, we have to hold quite a lot of information in our minds while we talk so we don't mind repetition. Giving more information than is strictly necessary is sometimes referred to as *redundancy*. For the same reason – the need to process information as we talk – we often prefer structures that are easy to unravel, for example, the use of 'and' as a connective, rather than more complex connectives such as 'while' or 'since'.

In writing, we try to avoid redundancy; we also try to vary the connectives that we use, so that, while 'and' is the norm in speech, in writing it is used only sparingly.

As speakers, we are also aware of the spontaneous nature or *immediacy* of speech, so we accept, and even expect, many structures that would be thought of as untidy or wrong in writing because we know that writers have plenty of time to edit and revise their language. For example, in speech we have false starts, complete changes of direction in mid-utterance, and utterances trailing off or not fitting together structurally, even though they make complete sense to the people in the conversation. David Crystal calls slips of the tongue, hesitations, back-tracking, and other features that result from the 'here and now' nature of speech as *normal non-fluency features*, and says that we are suspicious of speakers who do not use them – we rate such people as rather unnatural – too well-rehearsed to be genuine.

Speakers also use *fillers* to give themselves thinking time; *initiators* like 'well' or 'OK, then' or 'right' at the beginning of their utterances, to signal the fact that they are about to speak; and *vague completers* such as 'and that' or 'and everything' to round off what they have said. These structures are devices which help speakers give shape to their utterances.

None of these features would be considered appropriate in writing, unless we were writing a piece of dialogue that was trying to imitate real speech.

Another difference in the 'rules' we apply to speech and writing is that writing is expected to conform more to *Standard English*, while speech, especially in informal contexts, is allowed more *regional dialects*. This includes grammar, as well as vocabulary; so phrases like 'I ain't done nothing' and 'we was' are tolerated much more in speech than in writing.

A major difference between the nature of speech and writing is that speech is *interactive*. Many informal conversations involve fast exchanges between participants; but even where one speaker is holding the floor for some time, there is still an awareness of audience which affects the language structures used. For example, listeners are expected to offer *reinforcements* such as 'mm' and 'yeah', to reassure the speaker that what is being said is valuable; speakers often use *monitoring* features, such as 'you know?' and 'do you know what I mean?', to check that what they are saying is being attended to. Writing does not have this intimate give-and-take: even the fastest form of written exchange – the Internet chat-room – is still very slow in comparison.

The grammar of speech is also coloured by what we are using speech for at any particular time, and specific kinds of interactions will contain certain types of grammatical structures: for example, teaching situations are likely to contain many examples of questioning; children at play may well be trying to direct each other's behaviour, in which case they will be using commands; jokes and stories have their own particular shapes that speakers stick to as part of the genre. The differences between different types of *genres* of speech will be explored later.

ACTIVITY

Read through the conversation below and:

- identify some of the structures in the data that are characteristic of speech

- discuss the different types of questions used by the interviewer. Why does the first question elicit a very short response, while the second stimulates a much longer answer?

(Notes on this activity are on page 74.)

A = an A level student; B = a primary school pupil

A: what we want to do/is ask you some questions/about things that you've written/you wrote about um a space story/didn't you

B: yeah

A: what was that about

B: well um/we were doing about space/so I wrote a story about um/me and my friend/went up to space/and we um/and we saw um this alien/with um six eyes/and two mouths/and one nose/and um then/we um picked up some stones/to take back to England/and we um/we um drove out out in/we went out into space/and um we looked/and we and we didn't want/to go back to England/but um we stayed up there/for six days/then we all flew back to England/and then we showed the um rock/that we brought back.

ACTIVITY

Read through the material on page 52, which is the transcribed speech of a 79 year-old Lincolnshire speaker. Pick out:

- any structures which you consider to be examples of regional dialect grammar

- any structures which you consider to be typical of informal speech in general (but not dialect).

Make two lists for the features above. When you have finished, consider how far any of the structures you have picked out could be used in Standard English writing.

(Notes on this activity are on page 74.)

continued

SPEECH

well/I'm Harry Bruntlett/I live at number one Mill Lane/Louth/my
age is seventy-nine year old/now when I started school/I was at
Withcall/in those days/and we had to go right through them
cuttings/they was all chalk/you used to go with your shoes clen/in
a morning/and you got there/you thought you'd a pair of cricket
sandals on/by golly/it was a mess/now where did we go then/we
went/we left there/and we went to Kelstern/and I used to go to the
shop there/and fetch a loaf of bread for tuppence/and a bottle of
pop/it was tuppence/and the big bottles was fourpence/and they
had them glass alleys in them days/you had to put a stopper in/and
give it a thump/and you wanted to get the bottle/to your mouth/or
you'd lost half on it in gas/it had just gone all up/aye/and what else
was it/where did we go there/from Kelstern/on we went to
Lanscroft/now I had to go to work/when I was twelve/on a
Saturday/and (laughter) you didn't get a deal for it/neither/I'll tell
you/aye/twelve on a Saturday/it was/and I left school/when I was
fourteen/and I had to go to service/and I got to see to four
horses/and I got fourteen pounds a year/my overtime on harvest
time/was fourpence/and I/of course we come to Keddington
then/and er/so then we was in the Louth area then/we'd a gas
house in them days/we used to be leading coal/and making the gas
like/you could get a bag of coal for four/for er two shillings/so it
was/and Jacksons/they had no end of lorries of coal/in the coalyard
like/leading lorries of coal out/and there was Wilemans/and no end
on them/but by golly/things is altered/I can remember them
building Lacey Gardens/them houses at Lacey Gardens/I was a
boy/when it used to be a road/I can remember the men building
that/aye/I don't know

5 Spoken vocabularies

All speakers have a range or *repertoire* of different styles of speech from
which they select in specific situations. For monolingual speakers, these
different styles will involve changes in accent, grammar and vocabulary
within the same language; for bilingual or multilingual speakers, the
variations may well involve *mixing* aspects of their different languages
together, or *switching* completely from one language to another for a
considerable period of time.

Regardless of the number of languages speakers have in their
repertoires, their choices of style – and therefore vocabulary, as an
important ingredient of style – will be highly dependent on a number of
factors. Some of these are the following:

Status – The status of the person you are talking to will have an
influence on how formal a style you choose. For example, if you are
talking to a person who is higher status than yourself, you are likely to
become more formal. In this situation, your vocabulary is unlikely to
include such items as slang terms, intimate forms of address, swear
words, and vague expressions such as 'thingy' or 'whatsit'.

The number of people you are addressing will also be influential: in general, people become more formal as they address bigger groups.

Setting – A formal setting, such as an interview, will trigger more carefully chosen and formal vocabulary than, say, an informal chat with friends on a bus or in the pub.

Purpose – What you are using your language for will influence your vocabulary: if you are telling a humorous anecdote, for example, your vocabulary may well differ from that you would use in a serious debate or when giving a set of instructions.

Topic – Some topics are likely to make speakers more careful in their choice of words. We are particularly careful when we refer to taboo areas like death, sex, or bodily functions – although this doesn't necessarily mean that we become more formal. We may go the other way, and select terms that are considered 'impolite'. The fact that we may go to extremes in this way is an indication that the subject matter is uncomfortable or embarrassing to us.

Channel – The choice of channel, or medium – speech itself, as opposed to writing – is also likely to influence the style we use. Writing is, in general, a more formal channel than speech, if, as our definition of speech, we have 'informal conversation between friends in a relaxed setting'. Our most common interactions – spontaneous exchanges with partners, relatives, friends and acquaintances about the facts and experiences of day-to-day living – do not involve us in using a particularly formal style or wide-ranging vocabulary. However, it is dangerous to draw up generalisations which try to cover all forms of speech or writing, as there are many different types of each.

ACTIVITY

The aim of this activity is to explore and compare two different types of spoken interaction.

Compare the dialogue that follows in *Dialogue 1* with the telephone conversation you studied previously (on page 42) which, for the purpose of this activity, we will call *Dialogue 2*.

These dialogues differ according to channel (*Dialogue 1* is face-to-face; *Dialogue 2* is telephone discourse) and purpose (*Dialogue 1* is one speaker entertaining another via a playful description of an imagined scene; *Dialogue 2* is oriented towards problem-solving).

What differences in vocabulary and other features might result from differences in channel and purpose?

To compare these dialogues with a more formal spoken genre, look at the speech by Tony Blair on pages 63–64.

continued

DIALOGUE 1

This is a conversation between two 16 year-old students on the way home on the bus. They are discussing the forthcoming prize-giving event at their school. *...*= overlapping speech.

A: got a prize

B: GCSEs

A: oh yeah

B: as has most of the world

A: well

B: as you go up to the top/stand with your top two feet together/on the top step/so you don't fall/get hold of his hand/thank you very much/Professor Ashworth/now it's got/it's not just you say it once/but you've got well three years/hundred in each year/three hundred *girls*

A: *yeah*

B: quite a lot are getting two prizes/or something/so you sort of probably got five hundred kids/walking up to him/saying thank you very much Mist/I mean he'll get bloody bored/isn't he

A: (laughter)

B: standing there/just shaking all these hands/saying/having all these kids saying/thank you very much Mist/Professor Ashworth/I wouldn't mind/but no one can say it/Profesher Affworth

A: (laughter) and spit at him

6 The rules of interaction

Some aspects of speech which result from its interactive nature have already been suggested in this section: for example, the way intonation is used to signal the end of a speaker's turn (or the wish to continue); the use of deictics and ellipsis, reflecting the shared speech context; the use of particular structures connected with the need for speakers and listeners to register each other's reactions; the use of questioning as an interactive strategy; vocabulary and language choice which reflects the relationship between the participants, their shared setting, purpose, topic and channel. This last section aims to look particularly at how speakers and listeners organise turn-taking, and at some very general rules which exist in our minds about how conversations should be conducted.

Pragmatics

The linguist, H.P. Grice, in his influential essay 'Logic and Conversation' (in P. Cole and J. L. Morgan, 1975), tried to set out in very simple terms some of the 'rules' we take for granted in our interactions.

He said that conversation is a co-operative activity, where speakers work together for an agreed purpose, each person trying to help shape the conversation with his or her contribution. So one basic rule is that speakers are trying to be *helpful and co-operative*.

Here are some of the rules or *conversational maxims* which speakers try to follow:

1. *Maxim of Quantity*

Give as much information as is required, but no more than is required.

This is saying that we try to be as explicit and specific as possible.

2. *Maxim of Quality*

(a)   Do not say what you believe to be false.
(b)   Do not say that for which you lack adequate evidence.

This is saying that we try to tell the truth, and that we feel we should have some reasons for our opinions.

3. *Maxim of Relation*

Be relevant.

This says that we try to stick to the point.

4. *Maxim of Manner*

(a) Avoid obscurity.
(b) Avoid ambiguity.
(c) Avoid unnecessary wordiness.
(d) Be orderly.

These points relate to the way we talk – in other words, to our style. They suggest that we try to make our contributions clear, concise and well organised.

General

Grice claims that, in general, we aim for *economy* but *clarity*, so that we don't constantly have to stop the conversation to ask people what they meant.

If we are unclear about meaning, we go through the following procedure:

• We search the preceding conversation for some kind of connection

• If that fails, we search the physical context we are in

• If that fails, we go to our long-term memories

• If that fails, we ask!

Because we have a rule that conversations are co-operative forms of behaviour, we make lots of assumptions and 'read between the lines' when people speak. For example, in the following conversation:

A: I've run out of petrol. Is there a garage around here?
B: There's one up the road.

A assumes that B is indicating the garage up the road sells petrol, and is open.

Similarly, in the following exchange, A assumes B believes the corkscrew will be in one or other of the places mentioned:

A: Where's the corkscrew?
B: Try the drawer or the cupboard.

Grice calls this 'reading between the lines' *conversational implicature*. Implicature (or, as it is sometimes also called, *inference*) is not an obscure academic feature of language: on the contrary, it is at the core of many of our everyday interactions.

For example, the many notices we see around us often incorporate implicatures in their assumptions about our behaviour. In the following, can you see how the notices assume that we are behaving or have behaved in a certain way?

THANK YOU FOR NOT SMOKING AT THIS TABLE
THANK YOU FOR DRIVING CAREFULLY THROUGH OUR VILLAGE
THANK YOU FOR TAKING YOUR LITTER HOME WITH YOU

The operation of implicature, illustrated in these examples, falls within an area of linguistic study called *pragmatics*. This covers many different aspects that have in common the idea of embedded rules and assumptions in language: those things that are taken for granted by language users, and which often cause difficulty for foreign language learners.

One large area you could think about, if you are interested in exploring the pragmatics of language in more detail, is that of how politeness is encoded in English. For example, look at the two requests below. They both 'mean' the same thing, but one is framed much more politely than the other:

OPEN THE DOOR
DO YOU THINK YOU COULD OPEN THE DOOR, PLEASE?

Even a small addition in wording can alter the impact of an utterance. For example, look at the following examples, taken from supermarket checkout interactions. Try to explain why the second seems more polite than the first:

DO YOU HAVE A REWARD CARD?
DO YOU HAVE A REWARD CARD AT ALL?

Can you think of any more examples like these?

Another, perhaps more subtle, area you could explore is that where the surface form of an utterance doesn't match the underlying meaning, so that listeners know that they should 'read between the lines' in order to interpret the speaker's comment. An example of this is the following, where A and B are work colleagues and the venue is their workplace, late on a Friday evening. A is passing B's door and notices B at her desk:

A (to B): Are you still here?

This can't be a straightforward question, as B is obviously there. The question is the surface form of the language, but the underlying meaning is more like 'You are working very late' (or even 'You are working very late and I am, too').

Can you think of any more examples like this one?

ACTIVITY

Look at the descriptions of faults in the way people talk below. Try to match them to the maxims you have just been reading about: what rules are these people breaking?

a bore	a scatterbrain	a liar
a stirrer	a shifty person	a con-man
a gossip	a wind-bag	withdrawn
vague	elliptical	abstruse

Consider the following questions:

• Can you add to this list?

• Can you think of any people you know, who have the conversational faults above?

• Do you agree with Grice that we do have 'rules' for how we behave in conversations?

• Would you add further rules of your own, to describe our behaviour in interactions?

Some more rules

1 *Turn-taking*

Turn-taking is an important aspect of the way speakers co-operate in conversations. The basic rule of taking turns ensures that one speaker talks at a time, and that a change of speaker occurs.

The process of turn-taking involves active listenership as well as speakership. While speakers ask their listeners monitoring questions – such as 'do you know what I mean?' (see *The grammar of speech* page 50) – listeners give speakers a variety of signals about how the talk is being received. Reinforcements ('mm', 'yeah') given by listeners at regular intervals encourage speakers to continue; however, if reinforcements come too frequently, with a mid-level intonation, or if they come too infrequently, speakers become discouraged, and eventually dry up. Telephone conversations involve speakers and listeners in being particularly attentive to such signals, since other supports are absent. In face-to-face conversation, non-verbal behaviour (i.e. body language) plays a large part in turn-taking: for example, as speakers are drawing to a close in their turn, they often look directly at the person they are going to hand over to; at the same time, if listeners want to get a turn, they look directly at the speaker in order to signal this.

2 *Interruptions*

Listeners' reinforcements can take the form of enthusiastic overlaps, or longer utterances which interrupt the speaker completely for a while. Speakers know the difference between the kinds of intrusions which are the result of listeners' enthusiasm and other types of interruption which are attempts by listeners to take over the conversations and have a turn. Listeners may use their turn to offer a new perspective on the existing topic, or to start a new topic altogether.

3 *Topic changes*

Speakers engaged in conversation can range over many topics in a very short space of time. We do not require strong links between topics in speech – unlike writing, where ideas are often arranged thematically, in paragraphs, and readers expect paragraphs to be clearly linked. Topics may change as new speakers take over; but existing speakers can also move through a number of topics within one turn. Often, a speaker who is trying to get a listener to talk will try many different topics in order to find one that the listener will be able to take up.

ACTIVITY

Read through the conversation below, and try to make some sense of what was going on. The dialogue took place in a queue in a sandwich shop near to a sixth form college in Manchester.

The speakers are all students who know each other very well.

E = Emma   K = Kate   P = Philip   N = Nicola

... = overlapping speech

E: listen/right/I went to the toilet/and we went near this sand pit/right/and I got up the next morning/and all my shoes/were caked in white stuff/cement/or summat/oh god/what's it been like at college then/alright/I'm knackered/me/I'm dead tired/I can't stand this weather/it's too warm/*I*

K: *where's* Stephanie today

E: she'll be here/in a bit/we've got a lesson/I think/I hope she's got a lesson/anyway

P: (laughter) I had ice-cream/on there/(indicates record album he is holding) so I licked it off/because it was on *Madonna*

E: *wooh*/Philip

K: where did you go/this weekend

E: number one/on Friday night

K: we went on Saturday

E: I didn't see anyone I knew/on Friday night

K: no/most of them go/on Saturdays

E: I went in/and the bouncer/was all over me/wasn't he *Phil*

P: *mmm*/that bouncer/was well after *her*

E: *and* he went/and he/eh/we nearly got fuckin' chucked out/of this other club/'cause of you

N: *me*

E: *yeah*/callin' *everyone*

P: *(laughter)*

E: Stephs at it/I'm gonna kill Nicky

Now consider and make notes on the following questions, either in preparation for feedback to the whole group, or for your own individual file:

- Consider the various speakers' contributions in the light of Grice's 'conversational maxims'. Do any of the speakers break the rules, and, if so, how?

- How does the turn-taking procedure work in the conversation?

- Which of the examples of overlapping speech are reinforcements for the speaker to continue, and which are attempts to interrupt in order to get a turn?

- How many topics are attempted during the course of the conversation? How are changes of topic achieved?

- Consider some of the areas of spoken language you have studied in this section. For example, what part is played in the conversation by such aspects as: the speech context and non-verbal behaviour; deictics and ellipsis; use of regional dialect vocabulary and grammar; normal non-fluency features? What comments would you make about the vocabulary used in the dialogue? What triggering factors have influenced the participants' choice of language?

Genres of speech

Just as we use written language for a variety of purposes, so we have a range of spoken genres which are used in different situations for particular reasons.

This section aims to explore what speakers know about varieties of spoken text, and what their characteristics are.

ACTIVITY

Read through the pieces of spoken language below, and, for each, decide:

(a) where the utterance has been taken from, i.e. what genre of speech it represents

(b) how you knew the answer to point (a) above: what aspects of the language made it typical of a particular genre?

continued

SPOKEN LANGUAGE

1. I name this ship ...

2. We are gathered here today to celebrate the union of ...

3. Lords, Ladies and Gentlemen ...

4. You do not have to say anything. But it may harm your defence if ...

5. Have you heard the one about ...

6. Right, settle down now, today we're going to look at ...

7. I'm afraid I can't take your call at the moment, but if you'd like to leave a message ...

8. Good morning, the St. Clair Foundation, Diana speaking, can I help you? ...

9. Hello, are you looking for anything in particular, Madam?...

10. One potato, two potato, three potato, four ...

11. Listen, this really weird thing happened to me last week ...

12. Hi, Phil Drew here, taking you through all the way to midday ...

13. Hello, this is Janet here, calling at 11.15 on Tuesday ...

14. Dave, this is Mike from work. Mike, this is my partner, Dave ...

15. Madam Speaker, I should like to ask my honourable friend whether he ...

16. Space, the final frontier. This is the voyage of ...

17. Eyes down looking for a full house ...

18. Tinker, tailor, soldier, sailor, rich man ...

19. Can't beat the real thing ... can't beat the feeling!

20. We apologise for the late arrival of the 9.15 from Crewe ...

Were you surprised at:

- how quickly you were able to categorise the extracts above?
- how many different genres of speech there must be in use in society?
- how many spoken 'rituals' we have (e.g. certain set phrases and formulas that we expect in particular situations)? Can you think of any more spoken rituals?

Here are some ideas to get you started.

What do you say in the following situations:

- When you want to end a conversation on the phone?
- When you pick the phone up?
- When you want to get off the bus, and you need someone to let you out from your seat?
- When you walk into a restaurant where you booked a table for dinner?
- When you go into an Internet chat-room?

ACTIVITY

As linguists focus ever more closely on our everyday spoken exchanges, they are finding that many of these are considerably more complex than was ever thought. For example, Eggins and Slade (1997) offer the following template for the structure of 'gossip':

1. Speakers use third person pronouns throughout (i.e. he, she, it, they)

2. Speaker substantiates (i.e. offers evidence for) what is being described

3. (Optional stage) Description can be probed by listener

4. Subject being gossiped about is negatively evaluated

5. Wrap-up stage – conclusions drawn

Further optional stages during the above are:

(a) Defence by listener

(b) Concession by speaker

Think about your own behaviour in exchanges you might term 'gossip' (and that doesn't just apply to female readers, by the way!). Are the stages above accurate, in your experience?

ACTIVITY

Look again at the extracts from different genres on page 61.

Different spoken genres are sometimes called speech events: a speech event is a whole interaction with a specific purpose. Even a casual exchange of words at the bus stop with a stranger has a purpose: that of acknowledging that you are two people in a common situation, two human beings who are prepared to be friendly. The purely social, 'human contact' purpose of some of our interactions, where the contact itself is more important than the words, is often called *phatic communion*. This phrase was invented by an anthropologist called Bronislaw Malinowski.

How far can you categorise the extracts according to *purpose*?

Try to establish a purpose for each: what are people using the language for, in the various situations?

When you have finished, think about your own experiences of spoken language over the last 24 hours – the different *speech events* you have been involved in.

Make a list of some of the ways you have used spoken language during this time: what have you used speech for, and to whom? It will help you to tabulate some of your points. The details for one speech event have been filled in, to start you off:

Speech Events	Purpose	Participants
phone call (made by me)	apology (for not being able to meet up as planned)	myself and friend

ACTIVITY

Look in more detail at your notes on the *purposes* of the 20 extracts of spoken data. Were some much easier to categorise than others? If so, why?

Also analyse some of the different *speech events* you have been engaged in during the last two days.

Were there some types of interaction that cropped up more frequently than others?

What were the most and least common types of speech events to occur?

Why were some speech events more common than others, in your opinion?

ACTIVITY

What follows are some examples of different spoken genres, or speech events. Read through each one, and consider the questions attached. Make some notes either in preparation for feedback to the whole group, or for your own file.

1. Here is the opening of a speech made by Tony Blair to Labour party followers at the Royal Albert Hall on the night of the 1997 general election. By the time he was speaking, Tony Blair knew he was about to become Prime Minister, his party having been assured of a large majority.

   • How does Tony Blair use language to give a persuasive message?

   • How does the language of his speech differ from what you would find in spontaneous talk?

Thank you! Thank you! A new dawn has broken, has it not? And it is wonderful.

We always said that if we had the courage to change, then we could do it, and we did it. Let me say this to you: the British people have put their trust in us.

It is a moving and it is a humbling experience.

And the size of our likely majority now imposes a special sort of responsibility upon us. We have been elected as New Labour and we will govern as New Labour.

We were elected because as a party today we represent the whole of this nation.

And we will govern for the whole of this nation – every single part of it.

We will speak up for that decent, hard-working majority of the British people whose voice has been silent for too long in our political life.

And we'll set about doing the good, practical things that need to be done in this country – extending educational opportunity not to an elite

continued

but to all our children; modernising our welfare state; rebuilding our National Health Service as a proper National Health Service to serve the needs of Britain.

We will work with business to create that dynamic and enterprising economy we need.

And we will work with all our people for that just and decent society that the British people have wanted for so long.

2. Here is an extract from an English lesson involving a group of Year 7 pupils reading and discussing a poem about fog with their teacher.

• What aspects of spoken language suggest that the dialogue is from the teaching situation in general, and from English in particular?

• Explore particularly the teacher's use of questioning: why do his questions not elicit any lengthy responses from the pupils?

T = teacher;   P1 = Joanne;   P2 = James;   P3 = Frances

T: Frances/stand up/read out loud/(pupil reads poem) good/right/sit down/(teacher repeats the reading of the poem) right/Joanne/why does the fog move slowly/why slowly/go on

P1:   um

T: well what does slowly mean

P1:   um

T: why does the fog move slowly

P2:   because

T: go on/James

P2:   um

T: why slowly/why not quickly

P2:   because fog moves slow

T: because fog moves slowly/right/tell me/what does fog do/Frances/what does fog do to things

P3:   make things/like almost invisible

T: makes things invisible/good/then why do we say/hunch shouldered/what does that mean/if you see something/hunch shouldered/what does that mean/OK/what does that mean/Frances

P3:   hugged up/snugged up/ready to strike

T: ready to strike/that's good/why is it

P3:   well/cats always do/like hunch up/when they're going to pounce

T: good/good/a man called Eliot/wrote a poem/about fog/and he described it/as a cat being hunch shouldered/good/why then hunched/what does it mean/what does hunch shouldered/what kind of vision does it make/what kind of picture does it make/in your mind

continued

3. The following is a telephone conversation between an insurance agent and a caller phoning to arrange car insurance for himself, to cover the car he is going to buy.

- Why are there misunderstandings during the conversation? Try to pinpoint particular stages in the dialogue when confusions arise.

- Could you apply Grice's maxims of conversation (pages 55–56) to this dialogue, to identify what the speakers are doing wrong?

A = agent;   B = caller

A: right/who introduced you to us/sir

B: sorry/I beg your pardon

A: oh/you're going to have to speak up/I'm afraid/it's very noisy

B: Mrs Sheila Leone

A: oh/it's a name you're giving me

B: yes/that's the name/you asked who introduced me/to your your company

A: does she have a policy with us

B: yes/same car/but er/I'm buying the car from her

A: right/just find the source code then/sir

B: 45/Whitefield Gardens/Whiteacre Gardens

A: I/it's OK sir/I'm just/right/so it's for Mr

B: no/it's not for/that's the woman/Mrs er whatsername/Leone/that's the person

A: so it's for Mrs L/and her surname/sir

B: mine

A: no/the/who I

B: Le Leone

A: oh/that's her surname

B: that's/that's her surname/Sheila Leone

A: oh/I'm with you now/I'm sorry/sir

B: OK

A: S/Leone

B: Sheila Leone

A: L/e/o could you tell me/how to spell it/please

B: (talking to someone in the background) how do you spell/that Sheila Leone bit/L/e/o/n/e

A: n/e/thank you/right/and her address/please

B: address is 45

A: 45

B: Whiteacre Gardens

A: Whiteacre

B: yes

A: could/w/h/i/t/e/

B: w/h/i/t/e

A: a/c/r/e

continued

B: that's right a/c/r/e
A: Garden
B: Gardens
A: Gardens
B: South Norwood
A: South Norwood/is that Croydon/Surrey
B: yes/probably/I mean/I'm from Kent/so I wouldn't really know
A: South Norwood
B: yes/I think/just put it down/as South Nor
    Norwood/London/when when I address letters/South East 25
A: ah/so it's South Norwood/South Norwood/London
B: I think it's 25 South/but I'm not sure
A: London
B: South West/South East 25
A: South East 25/do you have a postcode
B: do I have a postcode
A: does she have a postcode
B: I don't know/love/I don't know
A: you don't know your postcode
B: no/it's not my postcode/it's her postcode
A: oh/um
B: these are the particulars/of the Sheila Leone
A: I see/yes/it's just that I need a postcode/you see/I have to/can I
    look it up/and ring you back
    (B talks to someone in the background)
B: 8/R/U
A: oh/sorry/8
B: yep
A: 8
B: R
A: R/for Robert
B: yes
A: U
B: yes
A: yes
B: that's right
A: is that it
B: that's it/yes
A: OK/do you have a telephone numb/a phone number for her
B: do I have a phone number for her
A: does she have a phone number
B: 892
A: so it's 0208/is it
B: double 2

continued

A: 0208

B: start again/I mean/you're fro/you're in London/aren't you

A: no/I'm not in London/sir

B: where you're phoning from now

A: no/I'm not in London

B: where are you phon/where am I phoning to now

A: Camberley

B: Cam where

A: Camberley

B: oh/so it'd be 0208

A: is it a London

B: would it be/would it be 0208/from you to London

A: yes

B: OK/put that 0208/892

A: yes

B: double 2

A: yes

B: 04

A: 04/thank you very much sir/right/and her present insurance/do you know when it expires

B: well/I don't know that either/I don't know that

A: oh

B: I'm not interested/I'm not interested in her insurance/I'm interested in getting a quote/for myself

A: for yourself

4. The following are three openings from oral anecdotes related by speakers who were asked to tell a story.

How are these beginnings typical of the way we start stories in speech?

(a) A while ago some friends of mine were sort of messing about with a ouija board ...

(b) When I was working at Norwood colliery there was er four men who were working along with me ...

(c) Last week Jeremy told me this story that his aunty er was coming home from the supermarket ...

5. The following texts are all attempts by writers to represent spoken language. Now that you have studied many different aspects of real speech, you will be better able to see strategies employed by the various writers below.

Read the texts and consider the strategies that the various writers have used to represent speech.

* How do the various representations differ from real spoken language?

* Why have the writers chosen to represent speech as they have?

continued

(a) In this extract from *Wuthering Heights* by Emily Brontë, Joseph, the
   Yorkshire manservant, has prepared some food for the young child,
   Linton. Linton has refused to eat it.

'Cannot ate it?' repeated Joseph, peering in Linton's face, and subduing
his voice to a whisper, for fear of being overheard. 'But Maister Hareton
nivir ate naught else, when he wer a little un; and what were gooid
eneugh for him's gooid eneugh for ye. Aw's rayther think!'

'I *shan't* eat it!' answered Linton snappishly. 'Take it away.'

Joseph snatched up the food indignantly, and brought it to us.

'Is there aught ails th' victuals?' he asked, thrusting the tray under
Heathcliffe's nose.

'What should ail them?' he said.

'Wah!' answered Joseph, 'yon dainty chap says he cannut ate 'em. But
Aw guess it's raight! His mother wer just soa – we wer a'most too
mucky to sow t'corn for makking her breead.'

(b) In this extract from *Tess of the D'Urbervilles* by Thomas Hardy, which
   is set in the West Country, Tess's mother has had a letter from their
   rich, distant relatives, the D'Urbervilles, offering Tess a job.

When she entered the house she perceived in a moment from her
mother's triumphant manner that something had occurred in the
interim.

'Oh yes; I know all about it! I told 'ee it would be all right, and now
'tis proved!'

'Since I've been away? What has?' said Tess rather wearily.

Her mother surveyed the girl up and down with arch approval, and
went on banteringly: 'So you've brought 'em round!'

'How do you know, mother?'

'I've had a letter.'

Tess then remembered that there would have been time for this.

'They say – Mrs. D'Urberville says – that she wants you to look after
a little fowl-farm which is her hobby. But this is only her artful way
of getting 'ee there without raising your hopes. She's going to own
'ee as kin – that's the meaning o't.'

'But I didn't see her.'

'You zid somebody, I suppose?'

'I saw her son.'

'And did he own 'ee?'

'Well – he called me Coz.'

continued

'An I knew it! Jacky – he called her Coz!' cried Joan to her husband. 'Well, he spoke to his mother, of course, and she do want 'ee there.'

'But I don't know that I am apt at tending fowls,' said the dubious Tess.

'Then I don't know who is apt. You've be'n born in the business, and brought up in it. They that be born in a business always know more about it than any 'prentice. Besides, that's only just a show of something for you to do, that you midn't feel beholden.'

(c) An extract from *Coronation Street*

Characters speaking:

T: Toyah Battersby, a schoolgirl who works part-time in the local cafe run by

R: Roy Cropper, a middle-aged educated man who is seen as a bit of a misfit.

As the scene opens, we know from a previous scene that Roy has been urging Toyah to hand in her school work, and has promised her that if she hands it in, she can have her wages early. Having money will enable Toyah to go and meet some new college-educated friends she would like to know better.

T: Right, I'm done.

R: Oh, right, er, you handed that project in?

T: 'Course. I could see when whatsisface took it he didn't rate it.

R: Whatsisface? Mr O'Neill.

T: Eh?

R: You never forget the name of a good teacher.

T: He didn't seem to be that impressed with it so I guess that'll be another 'D' then.

R: Do you care?

T: 'Course I do. I've told you, I'm fed up with people thinkin' I'm thick.

R: Why?

T: 'Cos, well, I get these ideas, right, but when I talk about 'em I make mistakes. See, I reckon half the stuff that goes on is rubbish and I wanna say so, but when I do, sometimes it don't sound right. D'you know what I mean?

R: Yeah, I know how that feels.

T: You don't, do you? I mean, you know loads o' stuff.

R: I suppose I do, but people think I'm a bit odd.

T: I don't, well, not *really* odd.

continued

R: Thanks, Toyah. No, you you see, you you you've got ideas and and opinions and you're curious and you get excited about things. I think you're on the right road. You just need to develop the skills to communicate that. It's people without ideas who are thick.

T: Yeah. but I've left it too late, an't I? I mean, the teachers think I'm useless, an' me mates, well, I can't let them see that I wanna learn owt.

R: It's not too late, Toyah. In fact, well I envy you.

T: Yeah, right.

R: Well, I do. I mean, you think I know things, well yeah probably I do, but what do I do with it? I run a cafe. You, you've got energy, you've got confidence, I mean, you're not shy, if you want to say something, you just say it. Well, some of us, we've got a lot to say, but it won't come out. Which would you rather be?

T: Yeah, but stuff I say, it it's rubbish.

R: If you want it to improve, then it will. You get ideas, don't you?

T: (nods)

R: Then you'll work out the answer. Here you are, try using this (hands over her wages).

T: Eh?

R: Work it out. You're not *thick*.

Answers and notes

Activity page 11

1. An information text where a piece of information is given and then expanded on in a number of 'bullet points'.
2. A letter.
3. A diagram where information is read in a clockwise direction; each piece of information relies on (i.e. is causally related to) the preceding piece.
4. This could be an ornamental plaque, e.g. a gravestone inscription or a concrete poem, where the shape made by the lines represents an object or idea in the real world.
5. A menu.
6. Information is read left to right; the boxes to the right in each case expand on those to the left.
7. A text where information is marked off in sections which are all equal in status (unlike No. 1, where the bullet points are equal to each other but secondary in status to the line at the top), e.g. the TV listings in a newspaper.
8. A poem.
9. A newspaper.
10. A tabulation.
11. A time-line, read from left to right, with information about each stage of the chronology.
12. Information set in a thematic structure: a summary of the subject in the centre, with subsections marked at radial points. The pieces of information on the 'spokes' do not rely on each other, as in No. 3.

Activity page 24   *Synonyms*

Anglo-Saxon/Viking Words (informal)	*French/Latin Words (formal)*	*Anglo-Saxon/Viking Words (informal)*	*French/Latin Words (formal)*
snag	impediment	cookery	cuisine
work	employment	date	appointment
smelly	odorous	talk	converse
put up with	endure	drink	imbibe
meet	encounter	underwear	lingerie
rude	offensive	job	occupation
stew	casserole	give out	disseminate
clothes/togs	attire	tuneful	melodious
kecks	trousers	start	commence
sweat	perspire	speed up	accelerate
living room	lounge	graveyard	cemetery
know-how	expertise	wash	launder
friendly	amicable	die	expire
worker	employee	booze	alcohol
house	residence	help	assistance
make	manufacture	meeting	rendezvous
keep	preserve	driver	chauffeur
task	assignment	drunk	intoxicated
loving	amorous	hair-do	coiffure

Activity page 30   *Film Review: 'This House Possessed'*

Graphology:
The graphological aspects of this extract are characteristic of the television page of a newspaper: the extract is in column form, with the title of the film in heavy type, in upper case lettering, and on a line of its own. The lines of the text are indented, and are relatively short, following the general convention of news columns.

Grammar:
The grammar of this extract deviates from the norm of Standard English usage in several respects. In the first sentence, the two adjectives 'spooky' and 'corny' follow the nouns that they modify, which is a reversal of the pattern that we expect. This pattern imitates that of the title of the film, for the purposes of humour. The humorous tone of the writing is reinforced by the way in which it approaches the informality of note-form in its grammar, particularly where the characters appearing in the film are listed. Minor sentences (sentences with no main verb) at the end give the impression that the writer is anxious to get to the end of the writing task: 'Hollywood veteran Joan Bennett as the old dear who knows what it's all about. Made in 1981'. The listing accompanies omission of the article 'a' to begin with ('sensitive nurse Sheila'; 'rock star Gary'), which suggests lack of individuality, and the use of 'the' later ('the cut-off couple'; 'the old dear') which suggests that the items are ingredients to be found universally, in every such film (the use of 'the' in this way is termed 'generic').

Modification is also used extensively, to imitate tabloid style: 'sensitive nurse Sheila'; 'rock star Gary'; 'Hollywood veteran Joan Bennett'; 'the cut-off couple'.

Phonology:
There is an extensive use of alliteration, which creates a playful, humorous effect: 'sensitive nurse Sheila'; 'mountain mansion'; 'series of sinister incidents'; 'cut-off couple'.

Semantics:
Many of the words and phrases used are colloquial, slang-like and sound out of date: for example, 'spooky', 'corny', 'old dear', 'knows what it's all about', the place is alive', 'cut-off couple'. 'Bad vibes' sounds like bygone 'hippy' language. The use of 'old dear' is particularly noticeable, carrying the connotation that the character is harmless but doddering – like the film itself. The worn-out vocabulary carries the message that the film plot is also very worn out, to the point of being comical without intending to be so.

Discourse:
All the language features noted add up to the fact that the writer finds the film particularly uninspiring – stereotyped, predictable, and dull. The choice of clichéd language, the use of alliteration, and the grammatical structures noted serve to send the film up by creating an ironic, tired-sounding style.

Poem: *Inventory*

Graphology:
The lines are very short and list-like; there is no punctuation except a full stop at the very end; one line stands alone at the beginning. There is a title, and an author's name. If the poem had conventional punctuation, there would be commas after each item: for example, nothing but nail parings, orange peel ... However, a list – such as a shopping list – would not have commas, necessarily. There is a link here with the title, 'Inventory'. The writer also obviously wants to omit punctuation in order to create opportunities to read some parts of the poem in more than one way.

Grammar:
The poem is one sentence, and is written in the form of a list of items, with descriptive terms attached to them. In some places, this directly imitates the structure of an 'Inventory', e.g. 'empty nutshells half filled' is like 'paint, two litres of' or 'lemons, fresh'.

In these cases, the modifiers would go after the nouns they modify – the opposite sequence to the norm for Standard English prose. The grammar is closely linked with the punctuation. The fact that there is no punctuation and the position of the line breaks mean that the modifiers could go either with the nouns before them or after them: for example, is it:

'empty nutshells half filled' or 'half filled ashtrays'?
'comb gap toothed' or 'gap toothed bookshelves'?

This helps to create a sense of uneasiness – the reader cannot pin down exactly which words should modify others; the writer wants to convey the uneasy, unsettled feeling that results from the departure of her lover.

Two parts of the poem could be read as sentences in their own right: 'you left me' and 'you shaped depression in my pillow' (if 'shaped' is taken as a verb). These depart from the idea of the list and make statements about the actions of the absent lover. They could be read as accusations.

Phonology:

There is extensive sound patterning in the poem via repetitions of various consonant groups and vowel sounds:

*n*othing	*n*ail			
*p*arings	*p*eel	*p*illow	sha*p*ed	de*p*ression
emp*ty*	dir*ty*			
nut*shells*	book*shelves*			
a*sh*trays	*sh*aped	depre*ss*ion		
le*ft*	hal*f*	*f*illed		
cups		night*caps*	comb	

The sound patterning helps to give a sense of cohesion to the writing, making it seem tightly knit, by setting up echoes; at the same time, full rhyme is avoided, because that would seem too neat and would contradict with the impression the poem wants to give of loneliness and incompleteness and loss.

Semantics:

The nouns chosen move between domestic items and aspects of physical bodies; what links both groups of nouns is that they all relate to waste and after-effects created by human beings in their lives together: parings, peel, dregs, ash, hair, impressions on items from the weight of bodies. This suggests decay and the passage of time; domestic items still show where humans have been, even when the humans are long gone. A sense of sadness is therefore created. At one point, the connection between the human body and an inanimate object becomes metaphorical: the bookshelves have gaps in their teeth where the departing lover has taken his/her books away. The sense of emptiness is further underlined by the fact that there are many references to containers which are no longer full, items that have been spoiled or outsides of things which are no longer attached to their centres: parings, peel, empty nutshells, half filled ashtrays, dregs of nightcaps ... The final 'you shaped depression' is the ultimate 'outside' of the departed lover, detached from the real person. The fact that the word 'depression' has been used both literally, to mean 'physical impression', and metaphorically, to mean 'feeling depressed', unites the two semantic strands of physical items and their psychological meaning or connotations.

Discourse:

The writer has employed all the levels of language to convey a message on two levels: a simple list of items which have been left behind by a departing lover; the sadness felt by the speaker at the loss of that person. The use of particular line breaks and the lack of punctuation cut across the reader's ability to read the poem as a simple list; the fact that the reader is unsure how items and descriptions of them relate to each other creates a feeling of uneasiness and uncertainty, mirroring how the speaker feels, and the phonological patterning creates subtle echoes of sound which reinforce the idea of incompleteness; the choice of particular nouns which suggest the idea of containers, the idea of waste and decay, and the idea of inside/outside contrasts, all allow the poem to maintain the theme of physical and psychological loss.

Activity page 35

(a) She 'ran to the *station/and 'caught the *train
(b) He 'fried the *onions/and 'chopped the to*matoes
(c) She 'went to *London/but 'wished she *hadn't
(d) They 'came in *early/and 'worked until *lunchtime
(e) He 'hurried a*long/to the 'centre of *town
(f) I 'got a *puncture/ 'while I was *driving
(g) 'After they *left/ 'I was ex*hausted
(h) She 'went to the *bank/at the 'end of the *road
(i) He 'hailed the *bus/and 'jumped *on
(j) Be'fore I *go/I must 'make a *phone call

Activity page 40

A: 'what do I `do/ with these ex `am papers
B: 'sort them into ˌbundles/'take them to the ˌclassrooms/and 'dish them `out
A: 'O`K/'what `else
B: 'that's about `it/I ˌthink/'how about using your own `initiative
A: 'that's ᵛ your job/'I'm just a ᵛ dogsbody

Activity page 49

Hairdressing

Activity page 51

Features characteristic of speech
    Initiator: well
    Filler: um
    Connectives: frequent use of 'and'
    Repetitions: drove out out; and we and we
    Change of course: out out in/we went out
    Utterances not fitting together structurally: so I wrote a story about um/me
    and my friend went up to space
    Particular use of 'this' for indefinite reference: this alien

Questioning
    The first question is a closed question: only certain responses are possible. In
    fact, because this particular question is a positive statement with a negative
    tag, the response 'yes' is strongly conditioned. Compare this with: It's a nice
    day, isn't it? It's rotten weather, isn't it? They're great shoes, aren't they?
    The second question is much more open, inviting the child to give a recount
    of the story, which invitation he takes up.

Activity page 51

Regional dialect grammar
    Demonstrative 'them': them cuttings; them glass alleys; them days; them
    houses
    Prepositions: in a morning; half on it; on a Saturday; on harvest time; no end
    on them
    Verb 'to be': they was all chalk; the big bottles was fourpence; we was in the
    Louth area; things is altered
    Present form of the verb used for past tense meaning: we come to
    Keddington
    Singular noun for plural: seventy-nine year old
    Contractions: we'd; you'd (for we had/you had rather than we would/you
    would)
    Negatives: You didn't get a deal for it neither
    Other phrases: what else was it; so it was
    Dialect vocabulary: aye; clen; a deal; alleys

Informal speech
    well; now; like (meaning 'as it were'); by golly; bottle of pop; I'll tell you; I
    don't know; to see to four horses; no end of lorries

4 Asking questions

Language is something we all know about and use; but it should be clear to you by now that, within language, there is a very wide range of different areas to study. You will also have realised that even within one area – such as spoken language – there are many specific and different aspects that can be looked at. In order to ask a question about language, you have to focus on a particular area to ask that question about. This is the situation that people are often in when they start thinking about a piece of research they might do.

Working back the other way, if you are given data to analyse (for example, in an examined paper), you need to bring to that situation some ideas about the appropriate areas to cover – the appropriate questions to ask – of the material that is being presented to you.

This unit will help you to think about:

- the theoretical areas language study can be divided into
- how different types of material illustrate those divisions
- the sorts of questions that can be posed about different areas of language.

As the unit proceeds, there is an increasing focus on preparing to do a piece of research; the next unit continues on the same basis.

However, even if you are not about to do an investigation, this unit and the next will still be useful to you, as both units contain extracts of language for you to analyse. In this case, imagine that these extracts have been presented to you as texts on an exam paper, and treat them as such: what would you say about them – how do they work?

On page 76 is a table which shows a possible way to divide up language study into different *language levels* and *language areas*.

LANGUAGE LEVELS are the different ingredients of language that you explored in the decoding sections of this book. They are as follows:

Phonology: the sound system of language
Graphology: the system of written symbols in language
Semantics: the system of meaning created by words and phrases used
Grammar: the system of language structures
Discourse: the way, whole language interactions or texts work.

	Language Acquisition (Picking up Language)	Language Change (Changes in Time)	Language Varieties (Differences)	Language and Society (Attitudes)
Phonological level	Acquisition of sounds	Changes in the sound system	Differences between sound systems	Attitudes to acquisition of, changes in, and differences between, sound systems and their uses
Graphological level	Acquisition of written symbols and other graphological conventions	Changes in the graphological conventions	Differences between graphological conventions	Attitudes to acquisition of, changes in, and differences between, graphological conventions and their uses
Semantic level	Acquisition of vocabulary and meanings	Changes in word meanings; invention and loss of terms	Differences between vocabulary used by or about individuals or groups	Attitudes to acquisition of, changes in, and differences between, vocabulary systems and their uses
Grammatical level	Acquisition of grammatical structures	Changes in grammatical structures	Differences between grammatical structures	Attitudes to acquisition of, changes in, and differences between, grammatical systems and their uses
Discourse level	Acquisition of a range of language functions (e.g. to entertain; to persuade) and spoken/written genres (e.g. oral narrative; written letter)	Changes in whole spoken/written genres	Differences between whole spoken/written genres; variations within a genre	Attitudes to acquisition of, changes in, and differences between, spoken/written genres and their uses

LANGUAGE AREAS are very broad subjects or topics which often form the basis of whole courses. They are as follows:

Language Acquisition: This covers the way in which we learn language, regardless of whether we are learning our first language or second, and regardless of what age we are. If we are learning the system of any language as part of the natural process of interacting with others in society, then we are acquiring it, or 'picking it up'; sometimes, the terms 'language learning' or 'language development' are used to refer to the more self-conscious process where language is not simply 'picked up', but consciously learned. Either way, we all acquire or learn something of the range of language levels outlined above.

Language Change: This covers the way in which language changes over the course of time: all aspects of language are subject to change, but the most noticeable aspect is change in word usage, with words dying out or changing their meanings, and new words being invented to describe new ideas and objects.

Language Varieties: This covers the ways in which language varies; according to such large-scale factors as region, age, gender, social class, and occupation, influencing the language use of whole groups of people for long periods of time; and also according to smaller-scale factors, such as the relative status of speakers or writers, their setting, the purpose and topic of the conversation or text, and whether speech or writing is being used. These factors influence the way individuals vary their language over relatively short periods of time.

Language and Society: This covers our attitudes to language, including such aspects as the status we give to particular groups and types of language; the way we stereotype others by their language use; and our various language 'taboos' which reflect the areas of experience we feel uncomfortable about.

ACTIVITY

Look at the data below.

What area on the table does each piece of language exemplify?

Try to match each example with at least one of the boxes. (You may decide that some of the examples could cover more than one box.)

1.
> little bonkey
> little bonkey on
> the busty road
> got to keep on
> Plobbind onhwarb
> with your Prestoys
> loaves

continued

2. bɑ bɑ bæk tip

(Baa baa black sheep)

3.

EVERYONE BATHES HERE AT THEIR OWN RISK

4. Simon's progress this year has been rather poor.

5. sjut (suit) ɔf (off)

6.

what ME?

If anything can keep him off the golf course
long enough to help dry up, it's a gay Old Bleach
pure linen cloth. Equip yourself with the new
heavier weight kitchen cloths as well as the original
rainbow-striped glass cloths. Both *ready for use*
— soft and absorbent without previous washing.

OLD BLEACH *ready to dry*
kitchen and glass cloths

OLD BLEACH LINEN CO LTD RANDALSTOWN N. IRELAND

continued

7. Fuss about 'fcuk' ads

IT DOES not matter how it is spelt, a swear word is still a swear word. The spat between French Connection and the Advertising Standards Authority over the use of the word "fcuk" in the fashion chain's advertising took a further twist yesterday when the ASA ruled that publishing the name in magazines for adults did not change its offensiveness.

French Connection had hoped to skirt an ASA ban on the use of "fcuk", which it says stands for French Connection UK, by putting dots between each letter and only publishing the original "fcuk fashion" adverts in magazines such as Elle and FHM.

The chain store argued that the grown-up nature of the magazines meant people would not be offended. But the ASA was not impressed. It has put out a "strong recommendation" to the Committee of Advertising Practice that its members should no longer take adverts for the chain. – *Kamal Ahmed*

© *The Guardian*

8. We was going down the road, like.

9. Ruth, aged 4, describing pins and needles in her legs: 'I've got fizzy legs.'

10. bæɵ (bath)

   kʌp (cup)

11.

Shamed By Your Mistakes In English?

Many people use such expressions as "Jane was invited as well as myself", and "was you going tonight?"

Still others say "between you and I" instead of "between you and me". It is astonishing how often "who" is used for "whom" and how frequently we hear such glaring mispronunciations as "tomorrER" and "reservOY". Few know whether to spell certain words with one or two "r's" or "m's" or with "ei" or "ie".

Indeed thousands of talented, intelligent people are held back at work and socially because their command of English does not equal their other abilities.

For example, most people do not realise how much they could influence others simply by speaking and writing with greater power, authority and precision. Whether you are presenting a report, training a child, fighting for a cause, making a sale, writing an essay, or asking for a rise . . . your success depends upon the words you use.

And now the right words are yours to command! A free book, "Good English — the Language of Success", tells you all about a remarkable, new home-study course which can give you a swift mastery of good English in just 20 minutes a day.

Never again need you fear those embarrassing mistakes. You can quickly and easily be shown how to ensure that everything you say and write is crisp, clear, *correct.*

This amazing self-training method will show you how to double your powers of self-expression, giving you added poise, self-confidence and greater personal effectiveness. You will discover how you can dominate each situation whether at home, at work, or even in casual conversation with new acquaintances. You will learn how to increase your vocabulary, speed up your reading, enhance your powers of conversation, and tremendously improve your grammar, writing and speaking — all in your spare time at home.

What's more, you will command the respect of people who matter, because you'll learn how to use English accurately, impressively, incisively — to cut through many barriers to social, academic or business success.

For your free copy of "Good English — the Language of Success", and proof that this unique home-study method really works, simply post the coupon on Page   — NOW. You have nothing to lose, not even a postage stamp, and you may gain a great deal by sending for this free book. For your own sake post the coupon TODAY. Or write to Practical English Programme, (Dept. IDE31), TMD FREEPOST, London WC2E 9BR.

continued

12.

Sprechen Sie German?
Not on one of our Handies

Ian Traynor in Bonn on a growing band of linguistic purists who are battling to banish anglicisms from everyday use

WHEN it comes to playing computer games, Germans do not reach for their *rollkugeleingabegerät* (roller-ball-entering-device). That mouthful, not so much a word as an "alphabetical procession", to cite Mark Twain on the wonders of the German language, is better known as a mouse.

German has sensibly appropriated the English term, banishing its own compound.

When it comes to the mobile phone, however, a more extreme case of language theft has been perpetrated – a step too far for the growing band of linguistic purists campaigning to keep German German.

The German word for the mobile phone is an English word that does not mean mobile phone – "Handy".

"We've developed this Anglicism our-

Mark Twain: 'German ought to be gently and reverently set aside among the dead languages, for only the dead have time to learn it'

selves," said Karin Frank-Cyrus, manager of the prestigious Society for the German Language in Wiesbaden.

"But many members are writing and calling to complain about the flood of Americanisms and Anglicisms. They are asking whether we can't use German words instead."

Spearheading the increasing clamour against the "rape of German by English", and against "Denglisch", the German equivalent of Franglais, is Professor Walter Kraemer of Dortmund University.

Two months ago the statistics professor and six colleagues founded the Club for

the Preservation of the German Language. It has 400 members and attracts 100 more a week, he says.

The Wiesbaden society has set up a commission to consider how to combat the tide of Anglicisms washing over German. It will report its findings in May.

"This is all about the colonisation of German by English," says Prof Kraemer. "The problem is getting much worse. Most Germans lack national pride and that really gets on my nerves."

A Deutsche Telekom phone bill, for instance, lists the number of "city calls" made. Similarly, if seeking information at the railway station, you go to the "service point".

Hollywood, transnational industry, advertising, the Internet, pop music – the cultural English onslaught is unremitting. Another German activist, the writer and educationalist Horst Hensel, says German children are "growing up today believing that songs are something essentially English".

© *The Guardian*

Check your findings against the answers on pages 86–87.

If you have been working in pairs, share your results with others in the whole group – how far do you agree on the areas covered by the language data?

ACTIVITY

Look again at the table on page 76.

A table like this can be useful in helping you to see how the different *language levels* could be the focus for a piece of research within the broad *language areas*.

Read through the list of Topic Outlines on page 82.

These outlines have been taken from A level Language investigations written by students during the last few years.

For each outline, decide which boxes on the table you would fit the outline into – what do you think the *main focus* was for each project? One project may well cover a number of different language levels, so don't feel you have to fit each outline into one box only.

In particular, you may find that you want to mark the discourse level as well as the smaller levels that go to make it up.

(Answers are given on page 87.)

If you have been working in groups, feed back your results to the group as a whole, while checking your answers.

How far are you in agreement about the areas and levels that would have been covered by the projects?

ACTIVITY

Now look again at the three Topic Outlines below.

If the outlines had been the versions in italics, what different levels or areas would the researchers have had to explore?

2. A comparison between the accent spoken in North Lincolnshire and Received Pronunciation.
    A comparison between the dialect of North Lincolnshire and Standard English.

13. A study of the techniques used in a range of contemporary magazine adverts for cars.
    A comparison of a number of car adverts – two from 1952 and two from 2000.

16. An analysis of a number of TV adult game shows, to establish whether there is a formula being followed.
    A comparison of two news programmes, one aimed at adults, and the other at children.

(Answers are given on page 88.)

TOPIC OUTLINES

1. A study of the pronunciation of an eight year-old boy who has been referred to the speech therapist.

2. A comparison between the accent spoken in North Lincolnshire and Received Pronunciation.

3. An examination of the use of swear words by groups of male and female students. I intend to explore the differences in usage, and to try to find reasons for the existence of any differences I discover.

4. An exploration of how language is used between counter staff and customers in food outlets.

5. A comparison of newspaper stories from *The Times* in the years 1920 and 2000.

6. A study of graffiti in different city centre venues.

7. An exploration of the spoken language of the Courts.

8. A study of the conversational rules in family interactions during meal-times.

9. A study of the language use of Afro-Caribbean speakers.

10. An examination of the development of children's narrative skills in writing; a comparison of stories written by seven year-olds and eleven year-olds.

11. A comparison between spoken and written instructions.

12. An analysis of a number of descriptions of sexual encounters in different literary texts.

13. A study of the techniques used in a range of contemporary magazine adverts for cars.

14. An analysis of the use and importance of personal names.

15. An examination of the way in which people stereotype others by their accents. I intend to record a number of differently-accented speakers, and ask informants to respond to the different accents.

16. An analysis of a number of TV adult game shows, to establish whether there is a formula being followed.

17. A comparison between 'Watch with Mother' (a children's TV programme from the 1950s and a modern children's programme.

18. Songs of Protest: an analysis of song lyrics by different artists aiming to register political protest.

19. A study of the language of gravestones, with particular reference to euphemisms about death.

20. An analysis of the language of crossword clues.

21. A comparison of problem pages from magazines written in 1939 and 2000.

22. An analysis of a range of political speeches given at Party Conferences Times.

ACTIVITY

Each group or individual should take one of the language areas you have been studying repeated below.

Draw a tree diagram which outlines some of the options possible in that area. For each area, a number of options have been given to help you; add any further options which come to you during the course of your work.

Try to give your tree diagram as many branches as you can, to cover all the options you can think of. If you decide you have exhausted your diagram, but can still think of more possibilities, draw some more diagrams. This activity is intended to be a brainstorm to give you ideas, rather than a neat and tidy exercise which will give you all the answers – so don't worry if your work ends up looking messy.

Language Area	*Options*
Language Acquisition	Adults or children? One individual or many? Typical development or not? Language features or functions? Speech or writing?
Language Change	Speech or writing? One language or more? Literary material or not?
Language Varieties	One individual or many? One social factor or more (age, gender, occupation, region, class, etc.)? Speech or writing? Literary material or not? One genre or a comparison?
Language and Society	Attitudes to which other language area (acquisition, change, varieties)? One level or more?

If you have been working in groups, each group in turn should pin up their tree diagram(s) and explain to the other groups the various options they discussed and drew for their language area.

When you have finished, consider the following questions:

- Have the activities you have completed so far helped you to understand more about the language areas and levels you could research?

- Have there been any aspects of the work you have done in this unit that have confused you?

Discuss these as a whole group if you are in a group situation; if you are working alone, make some notes on issues you would like to clarify, and discuss these at the next meeting with your supervisor.

Different types of question

The ordinary questions that you use in everyday conversation are also the questions that are relevant for research purposes.

Ordinary question words are: what; how; why; when; where; and who.

Whatever area or level of language you decide to research, the following are going to be relevant questions to ask:

WHAT? Researchers will always want to know 'What happens?' After getting a focus on an area to explore, a researcher will need to find some data, or raw material, to give him or her some evidence of what happens in that area.

HOW? A part of every investigation will be 'How does it work?' Whatever the research question, and whatever the data that has been collected, the researcher will want to know how the data 'works' – how it hangs together, whether there are patterns to be observed, what the 'rules' are.

WHY? The two questions above lead naturally to a third question – 'Why does it happen like this?' The researcher will always want to ask what the language tells us about the human beings behind it: what are their motives and concerns; what factors are working on them, to make them behave in this way?

Aside from the question above, you may also want to ask other questions, depending on the area of language being researched:

WHEN? When does/did this language use occur?

WHERE? Is this language use particular to a certain geographical or social area?

WHO? Are there certain individuals or groups who use this type of language?

ACTIVITY

In order to test out the usefulness of the questions you have just been reading about, read through the data on pages 85–86.

The data was collected as follows: *Menu A* is from a transport café on the outskirts of Manchester; *Menu B* is from a large chain hotel in Newcastle upon Tyne.

First, decide what your research question will be. Look back at the table of language areas and levels (page 76) to help you to decide the main area and levels you will be covering.

continued

Apply each of the questions: what? how? and why? in turn to the data, and write down some brief notes, either for your own file or for feedback to the whole group.

Then decide whether any of the questions: when? where? and who? are relevant to your study.

When you have finished, consider the following further questions:

- How useful were the questions: what? how? and why? Do you think these questions are always going to be relevant to language research?

- Which of the questions: when? where? and who? were relevant to what you were doing? Can you think of areas of language which would involve the questions that you didn't use?

MENU A

WAYFARER CAFE

ALL DAY FULL B/FAST WITH B/B

OR TOAST, INC POT TEA

HOMEMADE + 2 VEG, POTS

STEAK + KIDNEY, ALL IN

HOT-POT

HAM SHANK, CHIPS OR JACKET

LIVER + ONIONS

YORKSHIRES/MUSHY PEAS/BLACK PUD EXTRA

APPLE PIE + CUSTARD

BAKEWELL TART

PARKIN

VARIOUS SNACKS

SAUSAGE MUFFIN

BACON BARM

SCOLLOPS + GRAVY

CHIPS + GRAVY/CURRY SAUCE

OR TO ORDER

continued

MENU B

THE RAVENSCROFT SUITE
GOURMET DINNER DANCE

An interesting warm salad of smoked bacon, wild mushrooms and duck, quickly cooked and abound with a melange of winter leaves sprinkled in a walnut dressing

*

Peeled prawns bound in a tomato enhanced mayonnaise with diced pineapple and walnuts, nestled on a meli-melo of lettuce served in a glass

*

A collection of cured meats and poultry, nestled on a rustic salad and doused in a warm raspberry dressing

*

A terrine of fresh vegetables, sliced onto a coulis of tomato and fresh herbs

***

Supreme of fresh salmon attentively grilled, presented on a cushion of homemade noodles with a champagne sauce

***

Fillets of fresh monkfish spread with a mousse of scampi caressed in cabbage and poached, sliced onto a dry vermouth and avocado sauce

***

Medallions of pork pan-fried and masked in a pink peppercorn sauce accompanied by caramelized kumquats

***

Escalope of turkey folded with cranberry sauce, dusted in breadcrumbs and baked, escorted by a rich Madeira sauce

***

Rounds of venison quickly pan-fried and masked with a sharp blackcurrant sauce with just a suspicion of Juniper berry

***

A tournedos of beef topped with a liver parfait, enrobed in crepinette and oven-baked, served with a Madeira and truffle fondue

Answers

Activity page 77

1. Language Acquisition – Graphology

2. Language Acquisition – Phonology

3. Language Change – Grammar (Everyone ... their)

4. Language Varieties – Graphology/Semantics/Grammar/Discourse (Occupational Language)

5. Language Change – Phonology (Changes in the RP accent)

6. Language Change – Semantics ('gay Old Bleach')

7. Language and Society – Semantics/Graphology (Attitudes to taboo vocabulary and its spelling)

8. Language Varieties and Society (Dialect vocabulary and grammar and attitudes to this)
9. Language Acquisition – Semantics
10. Language Varieties – Phonology (Northern pronunciation)
11. Language and Society – Phonology/Semantics/Grammar (Attitudes to regional language and 'correctness')
12. Language Change and Language and Society – Semantics (Attitudes to changes in vocabulary)

Activity page 81

(N.B. Your decisions on levels may vary – in practice, this would depend on the precise nature of the data that had been collected.)

	Areas	*Possible Levels*
1.	Language Acquisition	Phonology
2.	Language Varieties	Phonology
3.	Language Varieties and Society	Semantics
4.	Language Varieties	Phonology, Semantics, Grammar, Discourse
5.	Language Change	Graphology, Semantics, Grammar, Discourse
6.	Language Varieties	Graphology, Semantics, Grammar, Discourse
7.	Language Varieties	Phonology, Semantics, Grammar, Discourse
8.	Language Varieties	Phonology, Semantics, Grammar, Discourse
9.	Language Varieties	Phonology, Semantics, Grammar
10.	Language Acquisition	Discourse (+ possibly Graphology, Semantics and Grammar)
11.	Language Varieties	All levels
12.	Language Varieties	Semantics, Grammar, Discourse
13.	Language Varieties	Graphology, Semantics, Grammar, Discourse
14.	Language Varieties and Society	Semantics
15.	Language and Society	Phonology
16.	Language Varieties	Discourse
17.	Language Change	Phonology, Semantics, Grammar, Discourse
18.	Language Varieties	Phonology, Semantics, Grammar, Discourse
19.	Language Varieties	Semantics as main focus
20.	Language Varieties	Graphology, Semantics, Grammar, Discourse
21.	Language Change	Graphology, Semantics, Grammar, Discourse
22.	Language Varieties	Discourse (+ Phonology, Semantics, Grammar)

Activity page 81   2.   Language Varieties – vocabulary and grammar, rather than phonology

13.   Language Change – same levels as before

16.   Language Varieties – the focus would be less on the 'rules' of the genre (discourse) and more on variations in vocabulary and grammar to accommodate the different audiences

5 Methods of data collection, organisation and transcription

This unit aims to get you thinking about the different ways you can collect language data, and the best way to present your data so that your reader is clear about what you were trying to do and what you have found.

Methods of data collection

A What is data?

Data is the raw material of a language investigation. It is the material you collect in your attempt to answer the question you have posed yourself in the original aim of your investigation.

Where you look for your data will be determined by your aim, but there are a variety of ways to collect data, and methods of data collection need careful consideration before you start. People whom you use as part of your investigation are called *informants*.

B Collecting spoken data

Here are some different ways to collect spoken language:

Note-making: This is where you write notes on what is being said.

Questionnaires on self-reported usage: This is where you ask individuals to tell you about the language that they use. You may ask informants verbally about their language, and write their responses down yourself, or you may simply give them a form to fill in on their own.

Tape-recordings with the participants' knowledge: Informants know that you are recording them, but you may tell them that you are looking for something other than your real aim. You may be recording a variety of different types of events – informants reading from word lists or set passages; informants producing monologues about particular subjects; interviews; a number of informants involved in a dialogue.

Tape-recordings without the participants' knowledge: Informants are not aware of being recorded; after you have obtained your recordings you are legally obliged to ask their permission to use the data. As above, you may be recording any one of many different situations.

Regardless of whether informants know they are being recorded or not, you, as speaker and researcher, have to make a decision about how far to be involved in any interaction.

Recordings taken from the media sources – for example, radio and TV programmes – also do not involve participant awareness of your particular recording, although obviously speakers were aware of the situation of the original recording or transmission. You, as a researcher, are not involved in any way in the interaction.

ACTIVITY

Consider the different methods described above, and, for each, write some notes on what you see as its strengths and weaknesses. Think about the practical aspects of collecting speech, as well as the possible warping factors involved in each method. Tabulate your ideas, using three columns:

Research Method   *Strengths*   *Weaknesses*

When you have finished, assess your ideas on the different methods described:

• Have you found clear patterns of strengths and weaknesses?

• Are there any further methods that have not been mentioned?

ACTIVITY

Go back to the list of Topic Outlines you considered in *Unit 4* (page 82).

Concentrate only on the investigations that involve the analysis of speech. For each of these investigations, say which method of data collection would be best.

Write some detailed notes on your choice of method, and any particular problems that you think could arise. Where investigations involve media texts, discuss how you would decide which programmes to record.

When you have finished, compare your results with those of the other groups, if you have been working in a group situation.

How far do you agree on the methods for each of the investigations on spoken language? Where you disagree, try to establish the issues involved which triggered your different choices. It may well be that different methods would work, depending on which particular aspects of language were being explored.

If you have been working alone, compare your ideas with the actual methods used (listed on page 96). Constructively criticise the actual methods used.

C Collecting written data

In some ways, written texts are much easier to collect than spoken ones, because of the practical problems involved in recording spoken language. But collecting written language needs careful preparation and attention too, in order to make sure that the data is suited to the question being asked, and in order to arrive at the right amount of material.

ACTIVITY

Look at the Topic Outlines again (page 82), this time focusing on those investigations that involve the analysis of writing, or where data could be drawn from both speech and writing.

For each investigation, decide what would be the important factors to consider in the collection of written material (or the spoken and written material).

For example:

- should the subject matter of the writing be kept constant for some of the investigations, and, if so, why? How can this be achieved?

- when the outline mentions a 'range', 'group', 'number', or just uses a plural and doesn't specify a number, how large a range should that be?

- should the items or people in the range or group differ from each other in some way?

When you have finished, pool your ideas on the important factors to consider when collecting written language, if you have been working in a group situation. If you have been working alone, compare your ideas with the factors outlined on page 96.

Data organisation

Readability

Any piece of language research is not simply an exploration for your own interest, although, obviously, that is an important element of your work. It should also, as a piece of communication in its own right, be accessible to any interested reader. For that reason, you need to give some thought to how your data should be presented and organised. This is not merely about where to place the raw data in your investigation; it is also about how your findings should appear. For this reason, issues about organisation should be considered at the outset and all the way through the process of research, not just at the final, 'fair copy' stage.

ACTIVITY

Look at the data below, from two people's answerphone recordings –
eight messages left on Steven's answerphone, four messages left on
Anne's. (All the real names and phone numbers in this data have been
changed.)

Imagine that your research is entitled 'The Language of Answerphones'
and that your aim is to explain to readers what some of the patterns are
in the way people leave their messages.

Devise a table in order to show some of these patterns. This means that
you will need to arrive at some suitable headings for the different parts
of the messages. For example, you might start with a heading on
'openings', and you might want to note how people introduce themselves
– whether they use their full names or not, whether they name the person
they are calling, and so on. Then you will need to think of other
headings that will enable you to show how the rest of the messages are
made up. Remember that the aim of this activity is to enable your reader
to look at the data you have collected in an economic and systematic
way. You are aiming to show connections between parts of the data that
are hard to see when the language is just written out in transcript form.
Your table is not intended to replace written analysis, but to support it by
enabling readers to see quickly what was going on in the messages.

When you have finished your tabulations, think about any further
dimensions you could explore if you had more data: for example, do
you think you could investigate language use according to the sex of the
speakers (and addressees)?

DATA

Steven

Hello it's me (.) are you not there yet (.) ring me when you get home

Steve this is a message from Peter Lawson (2) I'm phoning about (.)
to get an answer about the course and I need to know by Tuesday if
possible (.) can you get back to me before um seven tonight (.) I'm
going out then but I'll be in all of tomorrow (.) so I hope to hear from
you soon (.) bye

Hi Steven it's Louise (2) just phoned for a chat so if you want to um
(.) phone me back sometime (.) feel free

Steven it's um Richard Henman from Leeds (2) wondering if you can
call me back tomorrow maybe (2) the number's 01234 567890 (.)
thanks

Ah Steven it's Richard Henman calling from Leeds again (cough) just
wondering if you could call me back the number's 01234 567890
thanks very much

Hello Steve it's Nick just wondering if you're going swimming
tonight

continued

Steven it's Liz (2) I've got the arrangements for the wedding more or less sorted (.) the reception er (.) is still going to be on the Saturday night (.) but we're driving down there on the Thursday (.) if you still want to come down with us give me a ring (.) talk to you then (.) bye

Steven (2) John here (.) call me back when you get home (.) and I'll er expect to (.) hear from you later

Anne

Hi Anne it's Jane (2) um twent um twenty to seven on Thursday night (.) thanks for your message (.) and thanks very much for the little perfume (.) um I hope you have a nice Christmas and happy new year to you both (.) um (.) I'll probably try and ring you tomorrow if I don't speak to you later (.) so (.) um hopefully speak to you soon (.) hope to see you soon (.) OK then (.) bye

Hi Anne (2) um it's Jack (.) I'm just ringing about some work er (.) if you could give me a ring when you get home later (.) thanks a lot (.) um (.) bye

Hi Anne (.) it's Anne (.) it's er Wednesday about nine o'clock (.) and I've got the answerphone again (2) (laughing) and that's David in the background (.) can you give me a ring sometime please Anne (.) I'm dying for a chat (.) see you bye

Anne (2) it's Sharon (.) I er fed the cats (.) they didn't eat much (.) but I gave them the rest of that pork (.) you know and they just went absolutely mad (.) and er (2) I put them out because it it was a nice day (2) I hope you had a nice weekend (.) and I'll see you soon (.) bye

If you were working in groups, share your tables with the other groups when you have finished. It is not important how neat and tidy your tables are; the crucial aspect of this activity is thinking through how to be economical and clear in your data organisation. If you were working alone, show the data and your table to a reader, and ask him or her to respond to the following questions:

- Is it clear what main points you were trying to make in your tables?

- Which aspects of your tables did your reader(s) find helpful/unclear?

Then consider, for your own benefit:

- what the difference is between the information you have in your tables, and the sort of points you would make in your written analysis

- where you would put the spoken dialogues (i.e. your raw data) if this were really your investigation.

Data transcription

If you are attempting an investigation on spoken language, it is unlikely that you will want to transcribe all the material you record; your first task, then, is to *select* the parts that are most relevant to the question you are exploring. Your next decision is *how much detail to transcribe* in the data you have selected.

The purpose of transcription markings is to help the reader to recreate the spoken language in his or her own mind. A transcript should be free-standing and independent: in other words, a reader should not need to listen to the tape in order to know what is being referred to in the analysis. It is therefore important to mark those aspects of speech that are not obvious on the page. But what you decide to mark will depend on what aspect of speech your investigation is concerned with. For example, the transcript below (which appeared in *Unit 3*) is from an investigation which explored how people manage topic changes in conversations. For this reason, the researcher felt it was important to mark intonation, overlapping speech, and non-verbal behaviour, particularly eye contact.

ACTIVITY

Read the transcript aloud and act it out physically, attempting to recreate the speech situation by using the appropriate intonation and non-verbal behaviour. If you are working alone, ask some other readers to help you to realise the dialogue.

Can you get a sense of what the conversation was like?

The dialogue took place in a queue in a sandwich shop near to a sixth form college in Manchester. The speakers are all students who know each other very well.

E = Emma,   K = Kate,   P = Philip,   N = Nicola,
... = overlapping speech

E: (accelerando) `listen/ ,right/I ´went to the `toilet/and we ´went near this `sand pit/ ,right/and I ´got up the next `morning/and ´all my `shoes/were ´caked in ⱽwhite stuff/ce`ment/or ,summat/(rallentando, imitating tiredness) oh `god/ ´what's it been like at `college then/ ,alright/I'm `knackered/ ,me/I'm ´dead `tired/I can't ⱽstand this weather/it's too `warm/*I*

K: (looking directly at Emma) *´where's* `Stephanie today

E: (returning eye contact) she'll be `here/in a ,bit/we've ´got a `lesson/I ,think/I ⱽhope she's got a lesson/`anyway

(Philip looks towards Emma)

P: (laughter) I ´had ice-`cream/on ,there/(indicates record album he is holding) so I ´licked it `off/because it was on *Ma^donna*

E: (returning eye contact) *^wooh*/^Philip

continued

K: (looking directly at Emma) ´where did you ˋgo/´this week ˋend

   (Emma and Kate exchange eye contact while speaking to each other)

E: ´number ˋone/on ´Friday ˋnight

K: ´we went on ˋSaturday

E: I ´didn't see anyone I ˋknew/on Friday ˏnight

K: ˋno/ˋmost of them go/on ˅Saturdays

E: (looking at Phil) I ´went ˋin/and the ˅bouncer/was all ˋover me/ ˋwasn't he *Phil*

P: (looking at Kate) *ˆmmm*/´that ˅bouncer/was ´well ˋafter *her*

E: *and* he ˋwent/and he/(Emma suddenly notices Nicky standing behind them in the queue) ˋeh/we ´nearly got fuckin' chucked ˋout/of ´this other ˋclub/ˏcause of ˏyou

N: (looking round at her) *ˋme*

E: (looking directly at Nicola) *ˋyeah*/ˋcallin' *everyone*

P: *(laughter)*

E: (generally) ˋStephs at it/I'm gonna ˋkill Nicky

ACTIVITY

Look once again at the list of Topic Outlines (page 82), this time concentrating just on those that involve the analysis of spoken language.

Take each investigation in turn and make some notes on the particular aspects of spoken language that, in your opinion, should be marked in any data transcript.

It may be useful to refer to *Unit 3: Decoding spoken texts* (page 33) to help you.

Here is a list of areas that were covered in that unit:

- Intonation
- Phonemic transcription
- The speech context
- The grammar of speech
- Spoken vocabularies
- Interaction features

If you are working in a group situation, share your ideas on the types of transcription necessary for the investigations on speech. Where you differ in your results, how far is this because you had differing interpretations of what the investigations were trying to do?

If you are working alone, make some notes on your observations for your own use, and for discussion with your supervisor at an appropriate date.

Notes

Activity page 90   *Collecting spoken data*

Topic Outline (page 82)	Methods of data collection used
1	Tape-recording with participant's knowledge
2	As in No. 1
3	Questionnaires on self-reported usage
4	Tape-recording without participants' knowledge
7	Note-making (tape-recording in court is illegal)
8	As in No. 1
9	As in No. 1
15	As in No. 1, then questionnaires given to informants
16	Recording from different TV channels – three programmes
17	Recording from TV of one programme aimed at similar age group to 'Watch With Mother'
18	Three songs centring on specific issue – racism
22	Recording of two speeches – one Labour/one Conservative

Activity page 91   *Collecting written or spoken and written data*

Topic Outline (page 82)	Factors for consideration
5	Content needs to be the same or similar – three small stories covered by actual investigation, on theme of royal occasions
6	Venues need to be clearly different in nature – three venues used by actual investigation
10	Stories need to be on the same or similar subject – five stories chosen from each group by actual researcher
11	Instructions need to be for the same type of activity
12	Useful if different approaches taken by the texts – three texts used in investigation
13	Useful if texts show different techniques – four texts chosen by actual investigation
14	Range of data needed – newspapers plus questionnaires on self-reported attitudes used by actual investigation
19	Selection important, if want to concentrate on semantics as main focus
20	Two different crosswords used by actual investigation to show variety of techniques
21	Useful if magazine is kept the same – two problem pages chosen by actual investigation

Section B

BUILDING UP RESEARCH SKILLS

The aim of this section is to get you familiar in more detail with some of the language areas that can be explored for research purposes. Learning how to go about research is a skill that is acquired through practice. The units in this section are designed to give you that practice, so that, after working on them, you will be much more confident about undertaking your own independent enquiries. If you are not embarking on an investigation at the moment, the units that follow are still useful to you in a number of respects. For example, they cover a range of theoretical areas that are likely to come up on an examined paper, and analysing the data contained in each unit will give you some practice in responding to data-based questions in a systematic way.

Each unit covers a different language area, and does the following:

- Explains briefly what the language area is concerned with
- Offers a range of language data for analysis
- Outlines different questions that could be asked about the data
- Suggests ways to start working analytically
- Gives a list of different topics that could be attempted in that area.

1 Norms and variations

As the title suggests, this area is asking about options in language use. The word 'norm' refers to what happens, as a general rule; and 'variations' refers to alternative choices that can be made.

Studies on norms and variations always look at a system of language use; a small, often closed, area where it is possible to spell out the rules and conventions.

As a researcher, your work is to find out why certain choices of language were made – what are the triggering factors? What were people trying to do with their language, when they made the choices they did?

Examples: terms of address; graffiti; the language of gravestones; names of all kinds; book titles and 'blurbs'; the language of menus; greetings cards; colour terms; connotations of particular groups of terms, e.g. 'ladies' vs. 'girls' vs. 'women'.

The language of chocolate bars

ACTIVITY

On page 100 is a collection of chocolate bar names. These chocolate bars were all the confectionery on display in one sweet shop in a small town on 5 July, 2000.

Consider the names given to these sweets; what do you think the manufacturers were trying to suggest, in giving their product these names? What connotations do they have? Are there any patterns of usage here: can you group the names into categories, where the same or a similar idea is being suggested? Here are some possible headings to work with:

shape; texture; size/quantity/value; animal names; mythical figures; planetary references; lifestyle; ingredients; stamina/success in competition; exotic locations/cultures; private pleasure

Apart from the meanings of the words, are some names similar to others *graphologically* (the way the words are spelt, the way they look), *grammatically* (the type of word chosen, e.g. noun, adjective, verb) or *phonologically* (the sounds of the words when spoken)? Create some more groupings with these different language levels in mind.

continued

Ruffle	Marble	Maltesers
Drifter	Curly Wurly	Flyte
Wispa	Maverick	Lion
Galaxy	Toffee Crisp	Starburst
Boost	Ritz	Rolos
Turkish Delight	Astros	M&Ms
Skittles	Double Decker	Ripple
Picnic	Fuse	Spira
Munchies	Magic Stars	Dime
Caramac	Minstrels	Crunchie
Aero	Smarties	Caramel
Milky Bar	Twirl	Mars
Time Out	Bounty	
Milky Way	Flake	

When you have finished, do the same activities for the following chocolate box names

Biarritz	All Gold	Black Magic
Dairy Box	Milk Tray	
Moonlight	Roses	

ACTIVITY

Pool your categories and discuss your findings, if you were working in a group situation. Whatever your chosen method of working, consider the following:

- Were you able to find consistent patterns in the way these names work? In other words, can you see *systems of language* use in the data?

Looking at systems, and the options within them, is what this area of *norms and variations* is all about.

RESEARCH PATHWAYS

Savoury snack names (British people eat more savoury snacks per head than Americans); names for shampoo, perfume, aftershave, and other cosmetic products, possibly contrasting 'green' products such as those in the Body Shop with more traditional versions, or contrasting those aimed at men with those aimed at women; the names of household cleaners from different eras, e.g. soap powders, toilet products, scouring powders/creams.

The language of opticians' shop names

ACTIVITY

Below is a collection of names of opticians' shops. Read them through, and try to categorise these on the basis of the techniques being used. You may find the following headings useful as a starting point:

references to well-known idioms or sayings; use of homonyms (words that sound the same but have different spellings); use of numbers to replace words; words which suggest numerical order; polysemous words (where one word can have two different meanings)

First Sight	Look Right
Eye Openers	4 Sight
Second Vision	Vision Express
20/20 Vision	Eye Site
Eyes Right	4 Eyes
Spec Tackle	A Sight for Sore Eyes
Eyes Front	Eyeland
Eye 2 Eye	Spex
Eye Contact	Insight
Second Sight	New Look
Special Eyes	

When you have finished, try to think of some more possible names for opticians' shops. Here are some ideas, to get you going:

Popular sayings: an eye for an eye; giving someone the eye; if looks could kill; looking daggers; an eye for the main chance; eyeing someone up

Particular words/meanings: spectacular; a spectacle; a vision; visionary; hindsight; a sighting

ACTIVITY

Pool your categories and discuss your findings, if you were working in a group situation. Whatever your chosen method of working, consider the following:

- Were you able to group the names to show systems of language use?

- Decide why the companies used the particular techniques they did. What were they trying to suggest about their products?

Also consider your ideas on new names, explaining what connotations you were trying to attach to them, either orally or in note form for your file.

RESEARCH
PATHWAYS

Names of other high street shops – hairdressers, general household shops (e.g. 'Just Wot U Need', 'The Fulmonte', 'Top of the Pots'), clothes shops, beauty parlours, etc.

The language of greetings cards

ACTIVITY

Look at the two birth cards below and on page 103. Analyse the language used in the two cards, and how it differs on the basis of gender group.

Think about all the different levels of language at work in the cards.

- Graphological level (the card for the girl was pink, while the boy's card was blue)

- Phonological level

- Semantic level

- Grammatical level

- Discourse level

Reprinted with permission © Carlton Cards

ACTIVITY

If you were working in a group situation, share your findings on how the cards differ from each other, according to the gender of the baby. If you were working alone, write up your notes as a summary, in essay form.

Whatever your chosen working method, also consider the following:

- What other variations might you find, according to the type of audience being greeted (e.g. age)?

- How is the language used in these cards typical of greetings cards in general? What are the 'norms', or commonly found features, of such cards?

RESEARCH
PATHWAYS

Research:

- the variations in techniques found in greetings cards, e.g. use of humour; use of verse; interplay between the cover and the message inside; use of different images; variations in typeface/use of handwriting; formulaic expressions used for greetings; use of language involving particular connotations (e.g. archaic language, use of emotive terms)

continued

> • the range of events and messages we mark by the use of greetings cards. Do you have any idea whether other cultures mark the same types of events, or use the same sorts of technique in their cards?

Website

These days you can send e-greetings as well as the more traditional cards.

Investigate how far e-cards follow the patterns you have observed in the paper versions. Try the following, but note that there are many further sites you could explore:

http://www.hallmark.com

The language of 'Mills and Boon' book titles

ACTIVITY

Look at the collection of 'Mills and Boon' book titles below. Can you find any patterns or systems in the data?

Try to categorise the titles according to their semantic content – what connotations link several of the titles together, as trying to create similar pictures?

Create your own headings for the categories you find.

Are there also common grammatical patterns in the data?

Rage	Secret Fire
Fever	Dark Tyrant
The Caged Tiger	Loving in the Lion's Den
Desire	Dangerous Moonlight
Snow Bride	Untamed
With All My Worldly Goods	Kiss of a Tyrant
Lord of the Land	Wildfire Encounter
Midnight Lover	The Fires of Heaven
Dear Villain	King of Culla
Burning Obsession	Bridal Path
Night of Possession	Always the Boss
Summer in France	The Girl from Nowhere
Greek Island Magic	Makebelieve Marriage
A Modern Girl	Pacific Aphrodite
Dangerous Demon	Savage Surrender
A Girl Bewitched	Dangerous Compulsion
Sweet Conquest	Dear Demon

ACTIVITY

Share your ideas on the categories you have found, if you were working in a group situation. Whatever your chosen working method, consider the following:

- What do the various connotations and structures of the titles tell you about the world of 'Mills and Boon' books?

RESEARCH PATHWAYS

There are many further aspects of this type of romantic fiction you could explore, including the norms and variations in:

- the advertising 'blurbs' (and images) on the book jackets
- authors' names
- names of the heroes and heroines.

If you wanted to do some more extensive research involving audience, you could compare the 'Mills and Boon' titles with those aimed at younger readers.

The language of gravestones

ACTIVITY

On pages 106–107 is a list of headings, epitaphs and verses offered by a funeral director to help relatives to choose an inscription for a grave.

Read them through, and look for patterns in language use:

Semantics: Are there particular ideas that occur frequently?
What metaphorical uses of language are evident in the data?
What are the connotations of the words and phrases used?

Grammar: Why are there particular grammatical structures in the data that are not in everyday use?
What structural patterning is evident within some of the phrases?
Why is it used?

Phonology: What aspects of sound patterning are evident, and why are they used?

Discourse: What do your observations suggest about our attitudes to death? For example, what is the function of gravestones and the language we choose to write on them?
Who is the language written by, and who is it aimed at?
Why do we choose certain forms of language and not others?

continued

HEADINGS
   Cherished Memories of ...
   Precious Are the Memories of ...
   Our Lady of Lourdes Pray for the Soul of ...
   Treasured Memories of ...
   The Saviour Has My Treasure ...
   In Loving Memory of ...

EPITAPHS

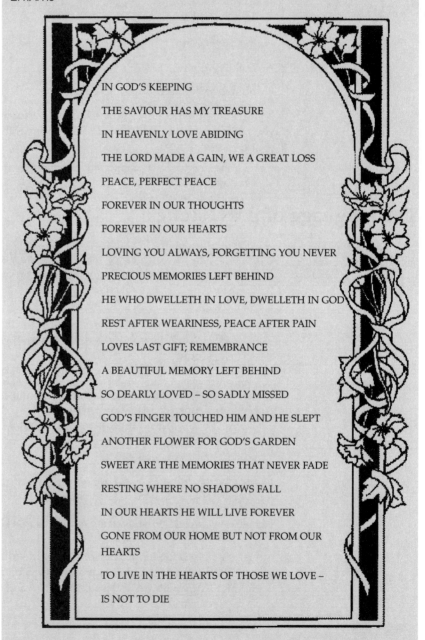

IN GOD'S KEEPING

THE SAVIOUR HAS MY TREASURE

IN HEAVENLY LOVE ABIDING

THE LORD MADE A GAIN, WE A GREAT LOSS

PEACE, PERFECT PEACE

FOREVER IN OUR THOUGHTS

FOREVER IN OUR HEARTS

LOVING YOU ALWAYS, FORGETTING YOU NEVER

PRECIOUS MEMORIES LEFT BEHIND

HE WHO DWELLETH IN LOVE, DWELLETH IN GOD

REST AFTER WEARINESS, PEACE AFTER PAIN

LOVES LAST GIFT; REMEMBRANCE

A BEAUTIFUL MEMORY LEFT BEHIND

SO DEARLY LOVED – SO SADLY MISSED

GOD'S FINGER TOUCHED HIM AND HE SLEPT

ANOTHER FLOWER FOR GOD'S GARDEN

SWEET ARE THE MEMORIES THAT NEVER FADE

RESTING WHERE NO SHADOWS FALL

IN OUR HEARTS HE WILL LIVE FOREVER

GONE FROM OUR HOME BUT NOT FROM OUR HEARTS

TO LIVE IN THE HEARTS OF THOSE WE LOVE –

IS NOT TO DIE

continued

VERSES

Nothing is more precious
Than the thoughts we have of you,
To us you were so special
God must have thought so too.

There's a place in my heart called memory lane,
And in it dear ... you will always remain.

Beautiful memories treasured forever
Of the golden days when we were together.

A tiny flower, lent not given
To bud on earth and bloom in heaven.

No one heard the footsteps
Of an angel drawing near,
As they took from earth to heaven
One we loved so dear.

Beautiful memories dearer than gold,
Of our loved one whose worth
Can never be told.

Memories are golden, we know that is true,
But we don't want memories we just want you.

A golden heart stopped beating,
Hard working hands laid to rest,
God broke our hearts to prove to us
He only takes the best.

ACTIVITY

If you were working in a group situation, pool your ideas on the language use on gravestones. Are there patterns of language use to be observed?

If you are working alone, write up your notes into a summary, in essay form. Consider whether you could tabulate any of your findings, for purposes of economy.

Whatever your chosen working method, consider the following:

- What are your conclusions about public attitudes to death, and the function of graves and gravestone language?

**RESEARCH
PATHWAYS**

Language use in newspaper obituaries; variations in gravestone inscriptions according to gender; analysis of the euphemisms (and dysphemisms – 'impolite' language) used about death; historical changes in the language on gravestones.

2 Accent and dialect

This area is concerned with the way in which language varies on a regional basis.

Accent refers to the sounds that are made by speakers; *dialect* refers to the words, grammatical structures and sayings that speakers have as part of their regional variety of language.

There is one accent, and one dialect, that are no longer regional varieties.

The accent called *Received Pronunciation*, while recognisable as originally a southern variety of speech, no longer marks someone as coming from a certain region. It now denotes someone as belonging to a middle-class social group, and is often the accent heard on national news broadcasts.

The dialect called *Standard English*, which originally came from an area between the Midlands and London, is now the agreed common standard language taught in schools and used in written material which is designed to be understood in all areas of the country. Standard English can be spoken with any regional accent.

In studies of accent and dialect, regional varieties are often compared with RP and Standard English as benchmarks against which to measure regional forms. This should not suggest that RP and Standard English are better forms of language than regional varieties – although the belief that this is so is an idea that itself can be explored in language research.

ACTIVITY

Look at the data on page 110, which is a phonemic transcription of an Aberdeen speaker's spontaneous speech.

Read it back to yourselves, deciding how the various words were spoken. You may need to refer to the phonemic alphabet which is given in *Section A, Unit 3, Decoding spoken texts* (page 33).

Then make a list of those words which are pronounced differently from how an RP speaker would say them.

continued

Remember that there are some features of connected speech which all speakers will exhibit – if you need to refresh your memories on this, look back to the section mentioned above, and make a list of the features to look for. Separate the ordinary features of connected speech from those uses of sound that are characteristics of the regional accent.

Finally, when you have your list of words, try to trace a pattern of sound usage in the Aberdeen speaker's language; for example, does a particular sound occur several times in the same context?

The speaker is a 40 year-old woman reminiscing about her school days.

In this extract, she is talking about how she did in a Geography test, and comparing herself with her school friend, Barbara.

Whole extract: I got three and she got two ... out of fifty. Aye ... that is absolutely ... and we were second last ... second bottom and bottom in the class, and that is, that is true, it ... it is. I wasn't a dunce ... I was a rebel ... I wasn't a dunce. Um, Barbara was more clever than me, I mean, like I say, she was ... she went into commercial ... or she could've gone into commercial, but she didn't.

I	got	three	and	she	got
aɪ	gɒʔ	ɵri	ən	ʃɪ	gɒʔ

two	out	of	fifty	Aye	that
tu	aʋʔ	ə	fɪftɪ	aɪ	ðaʔ

is	absolutely	and	we	were	second
ɪz	absəlutl̩	ən	wi	wɜr	sɛəkən

last	second	bottom	and	bottom	in	the
last	sɛəkən	bɔʔəm	əm	bɔʔəm	ən	ðə

class	and	that	is	that	is	true
klas	ən	ðaʔ	ɪz	ðaʔ	ɪz	tru

it	it	is	I	wasn't	a	dunce
ɪʔ	ɪʔ	ɪz	aɪ	wɔzn	ə	dʌns

I	was	a	rebel	I	wasn't
æ	wəz	ə	rɛəbəl	æ	wɔzn

a	dunce	um	Barbara	was	more
ə	dʌns	em	bærbræ	wəz	mɔr

clever	than	me	I	mean	like	I
klɛəvər	ðən	mi	ə	min	laɪk	ə

say	she	was	she	went	into
seɪ	ʃi	wəz	ʃi	went	ɪntə

commercial		or	she	could've	gone
kəmɛəʃəl		ɔr	ʃi	kudə	gɒn

into	commercial	but	she	didn't
ɪntə	kəmɛəʃəl	bʌʔ	ʃi	dɪdnə

ACTIVITY

Share your findings, if you were working in a group situation. Were you able to identify some patterns of sound which systematically differ from RP pronunciations?

If you were working alone, try to express your findings economically, by expressing them in tabulated form, or as a set of 'rules'.

Whatever your chosen methods of working, consider the following:

• Can you see that, although the speaker is using features of regional accent, she is speaking throughout in Standard English?

ACTIVITY

Read through the interview below, between a researcher (R) and a retired proprietor (P) of a small shop.

Try to identify some regional dialect features – vocabulary and grammatical structures – that differ from Standard English.

R: how did you keep your tick records

P: put it in a book

R: you haven't kept that book

P: no/I've had some good debts I've had some bad uns and I've had/I think about three years after I'd sold shop/I sold to a man what worked with me at tannery and he still kept on working/and I said to him how's business Bill/he said not so bad he said but you've some bloody bad payers haven't you/I said well no I haven't but that/that has gone out of existence/you see/we lived in Reservoir Street and there wasn't a shop in the district where they would let us have credit/in finish I'd got to go over to me Uncle Jim's in Dean Road on him to get us some bread butter cheese or anything

R: but quite a few people who came into your shop you let them have

P: oh yes yes/which was never very never very great/you'd get these as want cigarettes/want a pack of cigarettes

R: did they settle up with you at the end of the week or did they let it go on for longer than that

P: no the biggest majority settled at the end of the week/now they had trouble the next door where the mixed was/where they had the tea coffee sugar bacon butter cheese

R: why should they have more trouble than you

P: because theirs was eatables/the biggest majority was eatables/especially for the family where in those days they had/if you hadn't a family of four or five there was something wrong in that family/so/anyway/I told you we lived in Channing Street and there were/were it fourteen houses or is there sixteen houses on

continued

either side/only/one person was ill in that street/the remainder went in some time or another to see how she was going on/and if a baby was born/they were running in with basins of gruel/you know for years I never drank tea out of a cup or a pint pot/always had it in a basin/oh it's lovely out of a basin

R: tastes better does it

P: I'll tell thee/you'd go in and it was a common thing to see/the door wide open and about half-a-dozen home-made flour cakes stood up against the skirting board cooling off/where they did their own/where mother did her own baking

R: why would they cool them there/why

P: because it were cheaper/it was a common thing to see them coming back with a dozen of flour on their head and they were knocking it together/and/it was cheaper/they saved by baking their own/and I reckon in seven houses out of ten you would see on the table a stone jar containing black/treacle or syrup/syrup/treacle/or black treacle

Source: Manchester Metropolitan University Oral History Unit

ACTIVITY

If you were working in a group situation, share your findings on the interviewee's regional dialect features. If you were working alone, write a summary for your file of the main patterns of usage you found. Whatever your chosen working method, consider the following:

- It is a well-recorded fact that dialect vocabulary has been dying out for a long time, while accent and some aspects of regional grammar remain strong. Why do you think that is?

ACTIVITY

The student who researched the accent of the Aberdeen area also collected some Aberdeen dialect vocabulary and grammar by questionnaire method.

Consulting secondary source material, she offered a number of informants of different ages, sexes, and occupations the words in the column on the left, and asked them for a translation. Their answers were as follows:

clipe = a tell-tale (or to tell tales)

quine = a girl

loon = a boy

cowpe = overturn, turn over, fall over, make a mess

greetin' = crying, weeping

blake = shoe/boot polish

stottin' = drunk, bouncing (a ball)

continued

yurded = clarted(!), dirty, filthy

sappy dubs = mud, wet mud

cloot = cloth

humming = smelly, stinking

plooky = spotty, pimply

scunnered = fed up, sick of, been tricked

These terms were offered to informants as a stimulus to memory; they were then asked to volunteer some expressions of their own which they considered part of the Aberdeen dialect. Some of these expressions were as follows:

fit like = how are you?

fit a day afa wither = what a day – awful weather!

fits this fur wany wi? = what's this for anyway?

fit are ye deeing? = what are you doing?

fa? = who?

far i' yi gan? = where are you going?

muckle = much	bonny = good looking
bairns, geets = children	orra = odd, left over
stravaig = to stroll	sair heid = sore head
howk = dig out	redd = set in order
simmet = vest	chanty = toilet, chamber pot
starvin' = cold	skink, hough = shin of beef
ging = go	rake = to roam
hoose = house	kirk = church
wifie = woman	brig = bridge
al = old	gleg = sharp, quick
puggled, peched = tired	hippen = nappy
glaiket = silly, a show-off	the streen = last night
wee = little	ken = know
canny = careful	breeks = trousers
dinna = don't	sark = shirt
canna = can't	crabbit = bad tempered
grumph = a moaner	neeps = turnips
div = do	tatties = potatoes
feart = frightened	bampot = fool
scaffie = person who sweeps rubbish	gan hame = going home

continued

Now try to answer the following:

- The boundary between what we call 'accent' and what we call 'dialect' is not always clear-cut, because sometimes you could classify a term or phrase as either a variation of a pronunciation, or an entirely separate and different vocabulary item. Look at the list, and decide whether there are any terms you would call 'accent variation' rather than 'dialect vocabulary'.

- Which of the terms on the list would be known by many English speakers – even though they may not use them – and which would be entirely foreign? How do you think English speakers might have learnt some of the terms?

- Which examples would you call dialect vocabulary, and which dialect grammar?

- Try to group together the different terms as far as possible, according to the areas they describe or name. Are there dialect terms in the regional language of your own area or family which cover similar areas? Try to think of some equivalents.

- In the survey, many more examples were offered by older speakers than younger ones, and many more by working-class speakers than middle-class ones. However, everyone listed one or more of the first six phrases. How would you account for these results?

ACTIVITY

Share your conclusions on the questions outlined above, if you were working in a group situation. Whatever your chosen working method, do the following:

- Tabulate any examples you have of dialect terms of your own area. (For a list of common regional dialect grammar variations, see J. Milroy and L. Milroy (1993).)

- Consider the methods of data collection used by researchers in this unit. What are the strengths and weaknesses of each? Would it be valid to use a range of different methods in one research project?

RESEARCH PATHWAYS

Comparison between a regional accent and RP; comparison between a regional dialect and Standard English; exploration of the accent/dialect variations of one speaker in different situations; analysis of attitudes to accent variety (informants responding to differently accented speakers); analysis of accent use in TV advertisements; representations of accent and dialect in written material.

3 Stylistics

Stylistics is concerned with style choices in language.

Style is a rather wide and all-encompassing term which is often used rather loosely. In Linguistics, it refers, like the term 'discourse', to the way in which whole texts – spoken and written – work, on a number of different levels, to create an overall impression or message.

Stylistic analysis always asks two main questions:

- How is the text put together – how does it work? This may well involve consideration of how the text is distinctive: in other words, what makes it what it is?

- What are the effects of language use in the text – what is the text trying to do?

There is a virtually limitless list of texts which could be explored.

The following are just some which researchers have found interesting to analyse:
Examples: literature; newspapers; magazines; comics; advertisements; TV programmes; non-fiction school textbooks; particular genres of speech.

Stylistic analysis is much easier to attempt if you set up a comparison between different texts in the area you want to research. This is by no means the only way to approach stylistics, but it can be useful because language choices are easier to spot if you can contrast them with alternatives.

Literature

Literature is interesting to research because it is a very self-conscious and deliberate form of language use which aims to create specific effects.

Some forms of literature can be problematic, however, because they are too large and unwieldy to form the basis of a small and well-focused piece of language research. This is particularly true of prose fiction in the form of lengthy whole novels. What is needed, then, is careful consideration of the types of question that can be answered by looking at particular sections or extracts. This is the issue addressed by the activities in this unit.

ACTIVITY

Read the two extracts that follow. Both are from *Jane Eyre*, by Charlotte Brontë. *Extract A* is from the original text, published in 1848; *Extract B* is an abridged version of the same part of the story, designed for adult slow readers and published by Kennett books in the 1950s.

Two possible questions that could be asked about these texts are the following:

- What has been left out of the Kennett version, and why?

- How has the language used in the extracts changed over the course of time?

It would be possible in one project to ask both questions, as they are interrelated: modern editors would be anxious to avoid language that was archaic, in their attempt to simplify the abridged version.

Take each question in turn, and make some notes in preparation for feedback to the whole group, or for your own file.

EXTRACT A

I HAD forgotten to draw my curtain, which I usually did; and also to let down my window-blind. The consequence was that when the moon, which was full and bright (for the night was fine), came in her course to that space in the sky opposite my casement, and looked in at me through the unveiled panes, her glorious gaze roused me. Awaking in the dead of night, I opened my eyes on her disc – silver-white and crystal clear. It was beautiful, but too solemn: I half rose, and stretched my arm to draw the curtain.

Good God! What a cry!

The night – its silence – its rest, was rent in twain by a savage, a sharp, a shrilly sound, that ran from end to end of Thornfield Hall.

My pulse stopped; my heart stood still; my stretched arm was paralyzed. The cry died, and was not renewed. Indeed, whatever being uttered that fearful shriek could not soon repeat it: not the widest-winged condor on the Andes could, twice in succession, send out such a yell from the cloud shrouding his eyrie. The thing delivering such utterance must rest ere it could repeat the effort.

It came out of the third storey; for it passed overhead. And overhead – yes, in the room just above my chamber ceiling – I now heard a struggle: a deadly one it seemed from the noise; and a half smothered voice shouted –

"Help! help! help!" three times rapidly.

"Will no one come?" it cried; and then, while the staggering and stamping went on wildly, I distinguished, through plank and plaster:–

"Rochester! Rochester! for God's sake, come!"

continued

A chamber-door opened: some one ran, or rushed, along the gallery. Another step stamped on the flooring above, and something fell; and there was silence.

I had put on some clothes, though horror shook all my limbs. I issued from my apartment. The sleepers were all around: ejaculations, terrified murmurs, sounded in every room; door after door unclosed; one looked out, and another looked out; the gallery filled. Gentlemen and ladies alike had quitted their beds; and "Oh! what is it?" – "Who is hurt?" – "What has happened?" – "Fetch a light!" – "Is it fire?" – "Are there robbers?" – "Where shall we run?" was demanded confusedly on all hands. But for the moonlight they would have been in complete darkness. They ran to and fro; they crowded together, some sobbed, some stumbled: the confusion was inextricable.

"Where the devil is Rochester?" cried Colonel Dent. "I cannot find him in his bed."

"Here! here!" was shouted in return. "Be composed, all of you, I am coming."

And the door at the end of the gallery opened, and Mr Rochester advanced with a candle: he had just descended from the upper storey. One of the ladies ran to him directly: she seized his arm: it was Miss Ingram.

"What awful event has taken place?" said she. "Speak! let us know the worst at once!"

"But don't pull me down or strangle me," he replied: for the Misses Eshton were clinging about him now; and the two dowagers, in vast white wrappers, were bearing down on him like ships in full sail.

"All's right! – all's right!" he cried. "It's a mere rehearsal of 'Much Ado about Nothing'. Ladies, keep off; or I shall wax dangerous."

And dangerous he looked; his black eyes darted sparks. Calming himself by an effort, he added –

"A servant has had the nightmare; that is all. She's an excitable, nervous person: she construed her dream into an apparition, or something of that sort, no doubt; and has taken a fit with fright. Now, then, I must see you all back into your rooms; for, till the house is settled, she cannot be looked after. Gentlemen, have the goodness to set the ladies the example. Miss Ingram, I am sure you will not fail in evincing superiority to idle terrors. Amy and Louisa, return to your nests like a pair of doves, as you are. Mesdames" (to the dowagers), "you will take cold to a dead certainty, if you stay in this chill gallery any longer."

EXTRACT B

I had forgotten to draw my curtain, and when the moon rose, full and bright, its light roused me from sleep. Awaking in the dead of night I opened my eyes on her silver-white disc. I half rose and stretched my arm to draw the curtain.

Heavens! What a cry!

continued

The silence of the night was ripped apart by a fearful shriek. It ran from end to end of Thornfield Hall. My heart stood still; my stretched arm froze. The cry died, and sounded no more.

It had come out of the third storey. And now, overhead – yes, in the room just above my own – I heard the sounds of a struggle: a deadly one it seemed from the noise. A half-smothered voice shouted: "Help! help! help!" three times rapidly. And then: "Rochester! For God's sake, come!"

I heard a door open. Someone rushed along the gallery. Another step stamped on the flooring above and something fell; and there was silence.

I pulled on some clothes, though horror shook all my limbs. Everyone, it seemed, was awake. I went out into the gallery. Door after door opened. The gallery filled with ladies and gentlemen. "What is it?" – "What has happened?" – "Are there robbers?" – "Where shall we run?" was asked on every side.

"Where the devil is Rochester?" cried Colonel Dent. "I can't find him in his bed."

"Here! here!" came a shout. "Calm yourselves. I'm coming now."

The door at the end of the gallery opened. Mr Rochester appeared with a candle.

"It's all right!" he cried, and his black eyes darted sparks in the candlelight. "A servant has had a nightmare, that is all. She's an excitable person, and has taken a fit with fright. Now then, I must see you all back into your rooms, for, till the house is settled, she cannot be looked after.

ACTIVITY

Feed your results back to the whole group, if you have been working in a group situation. If you have been working individually, write up your notes using the following headings:

- What are the main omissions from the Kennett text, and why do you think the editors chose to abridge the text in this way?

- What aspects of language change are noticeable when the two texts are compared? In particular, are there specific words and phrases used in the original text that have died out or changed their meanings?

- Are there features in the original that may have been changed *either* to simplify *or* to modernise, and it is difficult to say which motive the editors had? Can you identify some examples of style that could be in either category?

It is important to realise that you do not need to have neat and tidy answers: the fact that you are able to speculate on the above questions is what language research projects are all about.

ACTIVITY

The text that follows is a whole short story, written by Virginia Woolf.

Its title suggests that it is a story written within a particular genre.

Before you start to read the story, brainstorm the ingredients you would expect to have in a story in that genre.

When you have finished, read through the story, and consider how far your expectations have been met: is Virginia Woolf following the genre, or is she playing with the reader's expectations?

A HAUNTED HOUSE

Whatever hour you woke there was a door shutting. From room to room they went, hand in hand, lifting here, opening there, making sure – a ghostly couple.

'Here we left it,' she said. And he added, 'Oh, but here too!' 'It's upstairs,' she murmured. 'And in the garden,' he whispered. 'Quietly,' they said, 'or we shall wake them.'

But it wasn't that you woke us. Oh, no. 'They're looking for it; they're drawing the curtain,' one might say, and so read on a page or two. 'Now they've found it,' one would be certain, stopping the pencil on the margin. And then, tired of reading, one might rise and see for oneself, the house all empty, the doors standing open, only the wood pigeons bubbling with content and the hum of the threshing machine sounding from the farm. 'What did I come in here for? What did I want to find?' My hands were empty. 'Perhaps it's upstairs then?' The apples were in the loft. And so down again, the garden still as ever, only the book had slipped into the grass.

But they had found it in the drawing-room. Not that one could ever see them. The window panes reflected apples, reflected roses; all the leaves were green in the glass. If they moved in the drawing-room the apple only turned its yellow side. Yet, the moment after, if the door was opened, spread about the floor, hung upon the walls, pendant from the ceiling – what? My hands were empty. The shadow of a thrush crossed the carpet; from the deepest wells of silence the wood pigeon drew its bubble of sound. 'Safe, safe, safe,' the pulse of the house beat softly. 'The treasure buried; the room ...' the pulse stopped short. Oh, was that the buried treasure?

A moment later the light had faded. Out in the garden then? But the trees spun darkness for a wandering beam of sun. So fine, so rare, coolly sunk beneath the surface the beam I sought always burnt behind the glass. Death was the glass; death was between us; coming to the woman first, hundreds of years ago, leaving the house, sealing all the windows; the rooms were darkened. He left it, left her, went North, went East, saw the stars turned into the Southern sky; sought the house, found it dropped beneath the Downs. 'Safe, safe, safe,' the pulse of the house beat gladly. 'The Treasure yours.'

continued

The wind roars up the avenue. Trees stoop and bend this way and that. Moonbeams splash and spill wildly in the rain. But the beam of the lamp falls straight from the window. The candle burns stiff and still. Wandering through the house, opening the windows, whispering not to wake us, the ghostly couple seek their joy.

'Here we slept,' she says. And he adds, 'Kisses without number.' 'Waking in the morning –' 'Silver between the trees –' 'Upstairs –' 'In the garden –' 'When summer came –' 'In winter snowtime –' The doors go shutting far in the distance, gently knocking like the pulse of a heart.

Nearer they come; cease at the doorway. The wind falls, the rain slides silver down the glass. Our eyes darken; we hear no steps beside us; we see no lady spread her ghostly cloak. His hands shield the lantern. 'Look,' he breathes. 'Sound asleep. Love upon their lips.'

Stooping, holding their silver lamp above us, long they look and deeply. Long they pause. The wind drives straightly; the flame stoops slightly. Wild beams of moonlight cross both floor and wall, and, meeting, stain the faces bent; the faces pondering; the faces that search the sleepers and seek their hidden joy.

'Safe, safe, safe,' the heart of the house beats proudly. 'Long years –' he sighs. 'Again you found me.' 'Here,' she murmurs, 'sleeping; in the garden reading; laughing, rolling apples in the loft. Here we left our treasure –' Stooping, their light lifts the lids upon my eyes. 'Safe! safe! safe!' the pulse of the house beats wildly. Waking, I cry 'Oh, is this your buried treasure? The light in the heart.'

When you have read the story and discussed how far Virginia Woolf has followed a well-known literary genre, there is another question that it's obvious to ask:

• Why is the story so confusing, and so difficult to make sense of?

We need to assume that Woolf was capable of writing clearly if she wanted to, so the answer can't be that she couldn't string a sentence together properly. So the confusion is deliberate. Why, then, did she write in this way, and how is the confusion created in the language? This could be the main question of a research project on this literary text. In order to answer these questions, think about the following:

• Why should the writer want to make the reader search for clarity and meaning in this text? Could that be part of the message of the story?

• How much confusion and deliberate suspense is created by:

   – the use of pronouns, referring to the various people in the story? This feature is called *reference* and is an aspect of *cohesion* in writing – in other words, the way the text 'hangs together'

   – unusual sentence structure, where important information is left right to the end?

continued

- sentences that have parts missing, e.g. objects, the thing or person having something done to it?

- *agency* – whether people or inanimate things are carrying out actions?

The passage also has a rhythmic quality which contributes to its pace and cohesion without clarifying any meaning. This means that, even though the reader might not be able to make sense of the story, he or she feels that the text has shape, and structure. This encourages him or her to struggle to solve the puzzle of the test.

In order to understand this process, look at the following features of language:

- Repeated grammatical structures, and variations in sentence length

- Sound patterning – the text's *phonological* structure.

ACTIVITY

If you were working in a group situation, share the results of your discussions on the various aspects of Virginia Woolf's story. If you have been working individually, write up your notes, using the headings outlined above.

Whatever your chosen working methods, consider the following overall questions:

- Can you see how the writing is very self-consciously making the reader search for meaning?

- Why should the writer want the reader to search for meaning?

Short stories and writing in particular literary genres often yield interesting research questions: short story writers have to pay particular attention to shape and structure because of the constraints of space; writing in particular genres usually follows a recognisable pattern. Both these aspects are of course traceable in the language used.

ACTIVITY

Below are three openings of stories: one is from a piece of autobiographical fiction; one is from a fairy story; and one is from a detective novel.

Decide which is which, and how you worked out the answers: how does the language used in such openings set up expectations in readers that they are going to get a particular type of story?

(a) Like most people I lived for a long time with my mother and father. My father liked to watch the wrestling, my mother liked to wrestle; it didn't matter what. She was in the white corner and that was that.

She hung out the largest sheets on the windiest days. She wanted the Mormons to knock on the door. At election time in a Labour mill town she put a picture of the Conservative candidate in the window.

She had never heard of mixed feelings. There were friends and there were enemies.

(b) Once there was a girl whose boyfriend drowned in the sea. Her parents could do nothing to console her. Nor did any of the other suitors interest her – she wanted the fellow who drowned and no-one else. Finally she took a chunk of blubber and carved it into the shape of her drowned boyfriend. Then she carved the boyfriend's face. It was a perfect likeness.

'Oh, if only he were real!' she thought.

She rubbed the blubber against her genitals, round and round, and suddenly it came alive.

(c) I had forgotten the smell. Even with the South Works on strike and Wisconsin Steel padlocked and rusting away, a pungent mist of chemicals streamed in through the engine vents. I turned off the car heater, but the stench – you couldn't call it air – slid through minute cracks in the Chevy's windows, burning my eyes and sinuses.

I followed Route 41 south. A couple of miles back it had been Lake Shore Drive, with Lake Michigan spewing foam against the rocks on the left, expensive high rises haughtily looking on from the right. At Seventy-ninth Street, the lake disappeared abruptly.

ACTIVITY

If you have been working in groups, share your results and discussions with the other groups. (Answers are on page 162, if you need them.) Whatever your chosen working method, consider the following:

- How did you work out which was which; how was the language of each characteristic of a particular genre of writing?

- What is the function of an opening of a story?

- What aspects are often covered in the way a novel or short story opens?

ACTIVITY

You could probably debate for some time the issue of what genre of writing the Bible represents – is it fiction or non-fiction, for example? Is it a recipe for good behaviour, or an instruction manual? Whatever the answers to these questions, it is often fruitful data for research purposes because so many different versions of it exist.

Below are two different versions of the 'Lord is My Shepherd' Psalm.

Analyse how they differ stylistically: what are the differences in the language used, and what effects do these differences create?

KING JAMES 1611

The Lord is my shepherd; I shall not want.

He maketh me to lie down in green pastures; he leadeth me beside the still waters.

He restoreth my soul; he leadeth me in the paths of righteousness for his name's sake.

Yea, though I walk through the valley of the shadow of death, I will fear no evil: for thou art with me; thy rod and thy staff they comfort me.

Thou preparest a table before me in the presence of mine enemies: thou anointest my head with oil; my cup runneth over.

Surely goodness and mercy shall follow me all the days of my life; and I will dwell in the house of the Lord forever.

YORKSHIRE DIALECT 1922

T'owd boss luks after mi, ah want for nowt.
'e sees as 'ow there's fields weer ah c'n sit missen dahn,
or tek a walk alongside o't' dams.

Bigod, ah feels missen agen,
's reight, t'foller 'im, 'cos e's t'Boss.

Aye, tho't' valley's thick wi' smooak,
an' Death's in it, ah'll fear nowt,
for th'art wi' mi, th'art mi backbooan an' mi
walking-stick; the keeps mi snug.

The ses: Sit thi dahn an' eeat;
ne'er mind as theer's them as grudge thee it,

continued

Tha smooths mi 'air, an' fills brimful mi mug.

F'shooa thi blessin' 'n compassion

'll be wi' mi till smooak clears
an' ah'm wi' thee
in t'ouse that's allus, allus thine.

ACTIVITY

If you have been working in groups, pool your ideas on the different versions of the Psalm: how do they differ in their language use? what are the effects of each text? why are the texts different?

Whatever your chosen working method, create some headings which would be useful to have as the basis for an analysis of these texts, then write up your notes under these headings.

RESEARCH
PATHWAYS

Differences between literary representations of speech, and real speech; descriptions of the same type of event in different pieces of literature, e.g. sexual encounters; analysis of fictive languages, e.g. in *1984* by George Orwell, *Riddley Walker* by Russell Hoban, and *Mother Tongue* by Suzette Hayden Elgin; male and female poets' treatment of the same theme; bias and techniques used in literature written for children; literature in translation compared with the original foreign language text; particular linguistic techniques used by an individual writer, e.g. Caryl Churchill, Harold Pinter.

Newspapers

Just as there is no one 'language of literature', so there is no such thing as 'the language of newspapers': although there are certain features which readers expect to see when they open their daily papers – such as photographs with explanatory captions, headlines, articles written in columns, and so on – the writers of news features choose language to construct the meanings that they want to convey on particular topics at particular times. So although it is possible to make some observations about certain conventions of language use in newspapers, there is no one language that they all follow.

Sometimes, contrasts are drawn between papers on the basis of their size, or on the basis of exclusiveness versus popular appeal; the larger 'broadsheets' such as *The Guardian* and *The Times*, which have relatively small circulation figures, are sometimes termed 'quality' papers, and are compared with the smaller, more popular 'tabloids' such as *The Sun* and

The Mirror. Such contrasts obscure a wide range of differences; for example, *The Sun* and *The Mirror* are polar opposites politically, as are *The Guardian* and *The Times*. The political affiliation of any newspaper is by far the strongest influence on what is reported in its pages, and how news is portrayed. All news reportage is mediated by the belief system of its management and editorial line, so there is arguably no such thing as 'the absolute truth': the idea, therefore, that 'quality' papers tell the truth while 'popular' papers perpetrate lies is, in itself, an example of bias.

Confused ideas about the social class of a paper's readership and about readers' 'intelligence' and reading ability also befuddle the analysis of newspaper language. Often, assumptions are made about these issues from the start, then the research becomes an attempt to make any 'findings' fall into line with what has already been decided: this is the opposite of research, since real research is an open-minded exploration with no narrow preconceptions about results.

So how can anything useful be discovered about language use in newspapers? The starting point must be language itself, and how it is used: all speculations about the implications of the language use need to come afterwards.

If you want to compare the treatment of a particular story in different newspapers, you need to choose newspapers whose political affiliations are different. An example of such a contrast would be *The Sun*, which is right-wing, and *The Guardian*, which is relatively left-wing. One difficulty in attempting to find a common story is that different newspapers tend to cover different topics. The exception to this is when a very big story breaks, which all papers will want to cover. However, since this is likely to be a major catastrophe of some kind – such as an earthquake, train crash, or ferry disaster – it is probable that all papers will treat the subject in a fairly dramatic way. News coverage in any one paper may extend to several pages, which then presents the problem of selection. It is better, therefore, to try to find a common story where the papers' different political stances are likely to come to the fore in the language chosen to describe it.

Another possibility is to go one step back, and analyse a story as it appears on a news agency 'wire', before it is bought and turned into copy for paper-based editions. Such an agency is the Press Association (PA) in the UK, which sells stories to many of the daily papers on a regular basis. The buying and selling of stories is an important aspect of the newspaper industry, and this process is now much more open to observation as a result of the use by news agencies of Internet websites.

ACTIVITY

Read the two PA articles that follow. Both were written by PA journalists, with a view to providing newspapers with different sorts of copy for them to buy. The central story concerns the conviction for multiple murder of a doctor from Hyde, Greater Manchester, on 31 January, 2000. The PA articles were written on that day, and the daily newspapers covered the story the day after.

Think about the following questions:

- How do the two articles differ in their treatment of the story?

- How do these two very different styles of writing reflect the types of stylistic difference found in paper-based newspaper stories?

ARTICLE A

SHIPMAN FOUND GUILTY OF MURDERING 15 PATIENTS

31 JAN 00

By PA News Reporter – Harriet Tolputt

Dr Harold Shipman was found guilty today of the murder of 15 of his patients.

Shipman, 54, killed the women between March 1995 and June 1998 while working at his surgery in Hyde, Greater Manchester.

A jury of seven men and five women at Preston Crown Court delivered the verdicts to a hushed courtroom on the sixth day of deliberations.

Shipman, who becomes one of the worst serial killers in modern British history, was found guilty of murdering Kathleen Grundy, 81, by injecting her with heroin on June 24, 1998.

In addition, he was convicted of forging widow Mrs Grundy's £386,000 will, an act which led to his downfall.

The jury also found Shipman, a married father of four from Roe Cross Green, Mottram, near Hyde, Greater Manchester, guilty of the murder of Bianka Pomfret, 49, Winifred Mellor, 73, Joan Melia, 73, Ivy Lomas, 63, Marie Quinn, 67, Irene Turner, 67, Jean Lilley, 59, Muriel Grimshaw, 76, Marie West, 81, Lizzie Adams, 77, Kathleen Wagstaff, 81, Norah Nuttall, 65, Pamela Hillier, 68, Maureen Ward, 57.

continued

Shipman had stockpiled huge quantities of morphine destined to relieve the pain of cancer patients.

He used the drug to kill his victims then falsified medical records to create bogus explanations for their sudden deaths.

He then bullied grieving relatives into believing there was no need for post mortem examinations on their loved one, which would have revealed his serial killing.

Reproduced courtesy of the Press Association

ARTICLE B

By Peter Beal and Harriet Tolputt, PA News

Sixteen times the word echoed round the oak-panelled number one courtroom at Preston Crown Court where the increasingly gaunt-looking GP, destined to become probably Britain's biggest serial killer, had spent 57 days on trial.

As the jury foreman, a middle-aged man in a grey shirt and tie, delivered the first "guilty" – to the charge that he murdered 81-year-old widow Kathleen Grundy – Harold Frederick Shipman barely flinched.

The 54-year-old doctor fixed his gaze on the far wall of the court some distance above the head of the judge Mr Justice Forbes as guilty verdict followed guilty verdict, each one repeated painstakingly by the woman court clerk.

Shipman, flanked on his left by two white-shirted prison officers with another on his right, moved his jaw as if muttering to himself as each of the verdicts which will ensure he dies behind bars for what the judge called "wicked, wicked crimes", came.

He wore the same muddy brown suit and nondescript striped tie that had been his dress for much of the trial that had started almost four months before.

With each day his hair had seemed thinner, his beard more grizzled and his cheeks more pronounced as he entered the dock.

As he waited for the verdict – which came almost unannounced just as relatives, journalists and members of the public gathered expecting the jury of seven men and five women to be sent home for a sixth night – the only signs of tension were the set of his jaw and a nervous fidgeting of the hands.

continued

The tension was clearly visible in the faces of the relatives of his victims, almost huddled together on one side of the public gallery at the rear of the court. It was to be replaced by gasps of relief and then tears as they realised the doctor was being convicted of the murder of each and every one of their loved ones.

A short burst of applause, quickly stifled, even broke out as the judge sentenced the GP to spend the rest of his life behind bars with some of the most damning condemnations heard in a British courtroom.

Mr Justice Forbes was clearly moved as he spoke to the relatives of Shipman's victims. He took the almost unprecedented step of removing his wig before telling them of his admiration for their courage and dignity, his voice trembling with the intensity of the moment.

The doctor's wife Primrose, 52, sat as far away from the relatives as she could be, on the other side of the public gallery, with three of the couple's four children, Christopher, 28, David, 20, and Sarah, 32. Their other son Sam, 17, stayed away.

Mrs Shipman sat with her head bowed as the verdicts and sentence was delivered, showing no apparent emotion. The family's mobile phone sounded during the judge's final words to the jury and the police in the case. Incongruously it played Beethoven's "Ode to Joy".

Shipman neither looked at his family, nor his judge, nor the jury who had convicted him as he was led from the dock with the judge's words that "you should spend the remainder of your days in prison" ringing in his ears.

Reproduced courtesy of the Press Association

ACTIVITY

Another useful approach to the study of language use in newspapers is to take a historical perspective, and trace the treatment of a particular subject over the course of time in specific papers.

On page 129 is a range of headlines all concerned with royal births, taken from *The Times* and *The Daily Mirror* in 1948 (the birth of Prince Charles) and 1982 (the birth of Prince William).

Decide whether, in your opinion, any changes have taken place in the way the two papers headlined the royal births.

Where changes have taken place, how would you describe the differences in language use, and how would you account for them?

THE TIMES

	1948	1982
Day 1 Page 1	A SON FOR THE PRINCESS	A SON FOR PRINCESS: 7LB BABY AND MOTHER DOING WELL
Page 2	CHEERING CROWDS AT BUCKINGHAM PALACE	CROWD MOBS NEW FATHER OUTSIDE HOSPITAL
Day 2 Page 1	WIDE REJOICING AT BIRTH OF THE PRINCE CONGRATULATIONS FROM ALL PARTS OF THE WORLD	CHEERS HAIL THE DEBUT OF A SLEEPING PRINCE

THE DAILY MIRROR

	1948	1982
Day 1 Page 1	AN 8LB BOY: BOTH DOING WELL	THAT'S OUR BOY
Page 2		NICE ONE CHARLIE ... NICE ONE SON
Day 2 Page 1	THE BABY HAS ELIZABETH'S FACE, HAIR LIKE PHILIP	FATHER'S PRIDE

ACTIVITY

Now look at the articles on pages 130–131, which are reviews of two football matches, one played on 1 January, 1900 and the other on 1 January, 2000, both involving Glossop, a Derbyshire team whose fortunes have clearly varied considerably!

(It might be useful for you to know that Glossop's local nickname is 'The Hillmen'; and that Mossley, a neighbouring town, are called 'The Lilywhites'.)

Both articles are from local papers, the Newcastle–Glossop match from the *Newcastle Evening Chronicle*, and the Glossop–Mossley match from the *Glossop Advertiser*.

- What differences in reporting style can you observe in these articles?

- These two pieces are both from a particular genre of newspaper article, the sports write-up. Are there aspects of this genre that have remained constant, despite the passage of time?

G CHRONICLE, MONDAY, JANUARY 1, 1900

TO-DAY'S FOOTBALL

LEAGUE: FIRST DIVISION.
NEWCASTLE UNITED
VERSUS
GLOSSOP

**At St James's Park,
Newcastle**

Positions
NEWCASTLE UNITED
Goal: Kingsley
Backs: Birnie and Gardner
Half-backs: Ghee, Aitken and Carr
Forwards: Rogers, Stevenson,
Peddie, McFarlane and Fraser

Forwards: Evans, Monks, Davidson,
Gallacher and Goddard
Half-backs: Clifford, Lupton and
Colville
Backs: McEwan and Rothwell
Goal: Williams
GLOSSOP
Referee: Mr J Fox

At St James's Park, Newcastle, Newcastle United and Glossop, in their meeting for League points, provided fare of an order acceptable to a holiday crowd. Looking at the lowly position, which the Glossopians occupied in the League table, it was not generally imagined that United had a very stupendous task on hand in bringing about their defeat, but it was just as well that a feeling of certainty was not generally prevalent, for the "certainties" very often prove disasters, and it is a notorious fact that the weak clubs have an especial "penchant" for lowering the flag of some team of superior calibre.

It was not forgotten neither, how Glossop, after being badly beaten by Aston Villa by 9 goals to nil, turned the tables later on their own ground, and defeated the Brummagem boys by a goal to nil.

Glossop have shown a great weakness in their matches on foreign grounds, but, as we have before remarked, it was just as well that the Newcastle United team had previously made up their minds not to hold their opponents too cheaply. Especially after their home win on Saturday, United were particularly anxious to improve their somewhat mediocre record.

The attendance was of considerable proportions, despite the counter attractions elsewhere, and notably at Sunderland. But the conditions were not of the best.

For one thing, a steamy haze hung over the field, and at the outset made it almost impossible to see what was doing in the far corners. Besides, the ground was hard and slippery. There had been a liberal sprinkling of sand to make the pitch more playable. Still, it was far from perfection.

There would be 15,000 people present, just before the start.

THE GAME

Newcastle were on this occasion fortunate in securing choice of ends, Davidson starting the leather for the visitors. Little could be seen of the work at the farther side of the field, the fog having rapidly got worse, but nevertheless it was seen that Glossop were pressing determinedly, and it was not surprising that three minutes after the start to hear shouts of "goal". It was quite true that

Monks had Beaten Kingsley

This early scoring was a good deal of a surprise, to those who could not see, particularly. The conditions, however, were as unfavourable to one side as the other.

Very soon after the ball had again been started from the centre, the Newcastle forwards were down at the Glossop goal. Here some infringement took place – of what nature we are unaware – but at all events Mr Fox was seen to place the ball on the penalty line, while the Glossop players came outside the same mark. It was at this point, the fog lifted for a moment, revealing Peddie preparing to take the kick. He made his drive, and, to the general consternation, he completely

Missed a Glorious Chance.

Glossop bustled up after this unexpected piece of luck, and again made inroads into Newcastle's ground. Monks just missing the mark. The visitors were then called upon to defend, and out of the fog came a roar which surely could only signalise a goal for Newcastle. When the players got near the Press-box, we were informed that such was the case and that

McFarlane Had Scored the Equaliser.

End to end the game raged, still leaving onlookers very much in the dark as to what was transpiring, but it was observed that Davidson made a good attempt to put the ball past Kingsley, which, however, he failed to do. Later, after a good save by Kingsley, Peddie was seen to be tearing down the field, evidently shaping for a goal. Rothwell, however, rushed up and kicked into touch. Looking at the position it was

A Well-Judged Save.

Further play occurred out of the range of our vision, but eventually it was noticed that Glossop were again in dangerous proximity to Kingsley, and that the Newcastle defenders were having their work cut out. Suddenly, once more the ominous shout arose from that portion of the crowd nearest Kingsley, and once more had it to be realised that, within 20 minutes

Glossop Had Scored Twice.

A deal of midfield play took place, and then again were the visitors the aggressors. Still, nothing could be seen of the work from our point of view, but evidently the Newcastle defence failed to check the persistent work of their opponents. By this time, we were becoming quite seasoned to the shouts of "goal" from Kingsley's end, so that the latest cry did not cause much surprise. It turned out to be Goddard, who

Scored Glossop's Third Goal.

There was a descent of fog, which obliterated the players' movements from our view, but no further score was made by either side, though most of the play was in the neighbourhood of Kingsley, the Newcastle custodian having to save on several occasions.

Half-time

Glossop ....3 goals
Newcastle United ....1 goal

The men were not long within doors, and the doings continued, as a wag in the Press box said, to be wrapped in "mistery." Shadow shapes were visible occasionally, and shouts near the goals indicated that something exciting was toward in the neighbourhood but as to what it was the reporters knew nothing. The proceedings suggested to some extent a shadow pantomime. While the ball was in midfield, the press men could catch a sight of the progress of events. It was here that

Gallacher Got Hurt

The player was taken "off," and hostilities were resumed with a fair amount of earnestness. What was puzzling most people at this stage was whether the game was to continue, and whether it was a League affair or otherwise. Enquiries made from those who might have known led to the expression of an opinion that

The Match Had to Count.

POINTS IN THE PLAY.

Monks scored for Glossop three minutes after the start.

Owing to the dense fog very little of the game could be seen.

Shortly after the re-start Newcastle were awarded a penalty kick. Peddie took the chance, and missed by yards.

McFarlane scored for Newcastle and equalised.

Gallacher put Glossop ahead.

Goddard scored another goal for Glossop.

Gallacher had to leave the field owing to injuries.

Peddie scored for Newcastle.

The game was stopped a quarter of an hour before time owing to the fog.

Final score

Glossop ....3 goals
Newcastle United ....2 goals

SHIPPING
INTELLIGENCE

(BY TELEGRAPH AND TELE-
PHONE FROM OUR OWN CORRE-
SPONDENTS).

Briton arrived Southampton. Dec 30
St Louis arrived New York, Dec 30
Mab arrived Buenos Ayres, Dec 30
Elfrida passed Beachy Head, Dec 31

LOCAL ARRIVALS AND
SAILINGS
THE TYNE

ARRIVALS, Dec 31 - Emily Rickett, Hartlepool; Hilding, Hartlepool; Ficaria, Copenhagen; Ardiethen, Cyprus via Hull; Loftus, Rosedale; Marsden, Moto, Consent, Sprightly, Gosforth, London - Jan 1 - Boldon, Swan, Beneficent, Eglantine, Albert, Hawthorns, London; Olaf, Hull, Dalston Hall, Leith, Mircelle, Fecamp; Edwd. Williams, Newhaven; Eturia, Grimsby; Durham, Invicta, London.

Christmas ghost in white returns to haunt Hillmen

Glossop 1 Mossley 2
by Syd White

A Bank Holiday crowd of over 300 turned out to see title contenders Mossley against relegation threatened Glossop.

And it was a ghost in white who returned to haunt the Hillmen.

Last season's top scorer Kevin O'Connell headed home the winning goal for the Lilywhites on 63 minutes.

Early on, ever present Glossop player Steve Reynolds burst through at pace on eight minutes only to see his left foot shot miss the target.

That, along with a Gary Thomas effort for Mossley on 16 minutes, was the only chance in the early stages, with neither team allowing their rivals to settle on the ball.

Eventually Mossley's patience paid off and their width and consistent shape began to pay dividends, with the mercurial Darrell Dickin seeing more of the ball.

Keeper Andy Merrick easily saved off the thigh of Thomas and then Dickin lifted his shot well over the bar.

But the Hillmen came back and on 36 minutes Simon Heaton linked up with Peter Sivori for Hamilton to cross to the far post where Wayne McGavin headed home to make it two in two weeks.

The lead was short lived as Mike Wolstenholme got on the end of a cross to put the Lilywhites level minutes later which left the game all square at the interval.

Mossley took a deserved three points as they had a majority of the second half – they made it 2-1 in the 68th minute when O'Connell headed home.

Ex-Mossley players Phil Hulmes and Dave Blow attempted to install some passion back into the Hillmen.

On 77 minutes Andy Levendis spurned a glorious opportunity to level but it wasn't to be as the Hillmen lost the first of the Christmas double header against their Tameside neighbours.

Glossop Advertiser

RESEARCH PATHWAYS

Further historical comparisons of similar events, e.g. the Titanic/Zeebrugge disasters; study of particular news features, e.g. weather forecasts; study of how men and women are described in the press; contemporary news coverage of a particular event; study of a specific language level and its effects across a range of papers, e.g. phonological patterning, use of puns/metaphors in headlines and articles in the tabloid press.

Websites

Most newspapers have their own websites, for example:

http://www.the-sun.co.uk
http://www.independent.co.uk
http://www.mirror.co.uk
http://www.guardian.co.uk

but you will find that the online texts are often different from their paper-based versions. This, of course, could be a research topic in itself.

Local papers often have e-versions, too. They are often linked into sites that advertise the local area: for example, the *Manchester Evening News* can be found at Manchesteronline.co.uk

For your own area, try typing in the name of your village, town or city, followed by .com

To look at PA stories, go to the PA site at:

http://www.pa.press.net

You can search for stories using the 'Storyfinder' button.

Magazines

The magazine trade is a huge industry which caters for a wide variety of different groups. To get you started on thinking about the nature of magazines and the way they vary, try the activities below:

ACTIVITY

What are the purposes of magazines?

Here are some suggestions. Try to think of one example of a magazine which has the purpose outlined, and add any further purposes you think are relevant:

To inform   To instruct
To entertain   To persuade

How do you think magazines differ from newspapers in the functions they fulfil?

ACTIVITY

How do the target groups for different magazines vary?

Here are some possible variations in audience.

Can you think of any magazines that fit the variations outlined?

Add any further variations which you think are relevant:

Different age groups   Different religious groups
Different gender groups   Different ethnic groups
Different interest groups   Different political groups
Different social class groups

ACTIVITY

The word 'magazine' is an Arabic word (originally 'makhasin'), meaning 'storehouses'. It is still used to mean a store for explosives or a supply-chamber in a machine (e.g. a magazine of rounds of bullets in an automatic gun).

Can you see any link between the original meaning of this Arabic word and the idea of a written publication? Does the Arabic word give a clue to the nature of the modern magazine?

Brainstorm the typical ingredients for one type or 'genre' of magazine: what features would a reader expect to find in it?

ACTIVITY

Consider the following:

• Were you able to identify magazines written for different purposes, or do most magazines fulfil more than one purpose?

• Were you able to identify different target groups for a range of publications?

• How far were you able to list the typical ingredients of your chosen magazine genre?

The activities you have just worked on should have helped you to see the complexity of the area of written language we refer to as 'magazines'. To try to analyse a whole range of magazines, or even one magazine, without having a clearly defined question in mind, and without a sharp focus on particular features, will mean that you won't be able to say very much of interest or value about the language.

In order to analyse the language, it is often useful to set up a contrast of some kind between different magazines with a focus on a particular feature. Then you can speculate, on the basis of proper evidence, about the purposes and audience for the language you have analysed.

To see what this means in practice, try the activity below:

ACTIVITY

Look at the three 'problem pages' on pages 136–138. Read them thoroughly. They are all from *Woman* – one each from 1947, 1987 and 2000.

The problems dealt with on the first two pages have some areas in common.

Make notes of the following language levels. Some examples of what to look for in each area have been given, to help you:

- *Graphological level:* How do the layouts, use of images, typefaces, etc. vary, and what are the reasons for the variations?

- *Semantic level:* What are the connotations of the different names of the agony aunts? How do their names, and the way they are described, represent their roles? What differences in vocabulary are noticeable in the three texts? Look particularly at the language used to describe behaviour.

  Do the agony aunts differ in how they address the writers of the letters (and therefore the readers of the page, by implication)?

  Do the three texts show any differences in levels of formality of language use? If so, can you find some examples of language to illustrate this?

- *Grammatical level:* How are statements and questions used to express meaning in the three texts?

  Are there any differences in the degree of certainty or hesitancy expressed by the three agony aunts?

  Are there any differences in how personal or impersonal they sound?

- *Discourse level:* Are there any differences between the whole purpose of the three texts? For example, are they all simply a direct question-and-answer correspondence?

  What does the language of the three texts tell you about any changes in society's attitude to relationships and behaviour?

  Have there been any changes in how agony aunts are seen, in *Woman* magazine?

ACTIVITY

If you have been working in groups, share your ideas on the three problem pages.

Whatever your chosen working method, write up your notes in summary form under the headings given. Also consider the following:

*   Do you feel that a comparison such as this can enable a detailed focus on language: how much detail were you able to go into, in your analysis of these texts?

*   What advantage is there in having some of the subject matter in common in the three texts?

ACTIVITY

Taking a historical perspective on a particular feature is one way to achieve a good focus on language in projects on magazines.

Another possibility is to look at a particular feature or features across different magazines published at the same time but aimed, either at different audiences (e.g. teenagers vs. adults), or at the same type of audience but with slightly different beliefs and concerns (e.g. women of different social classes). In this case, you would be looking at possible variations according to some of the groups you brainstormed in the activity on page 133.

To get an idea of what this might entail, look at another problem page on page 139 – this time from *Woman and Home*, a contemporary of the 1947 *Woman* magazine on page 136.

How does the *Woman and Home* text compare with the *Woman* problem page you have just been studying? Analyse the two texts and make some notes as before.

ASK Evelyn Home

If you have a personal problem, there's a shadow over the sun. Let Evelyn Home help to smooth away the trouble —her address is c/o WOMAN, 186 High Holborn, London, WC1. Send a stamped addressed envelope for her reply

SEEING HIS WIFE

LOVE can be a cruel emotion, as cruel as it is strong. Its cruelty lies in its blindness—people in love tend to see only themselves, their eyes are closed to whatever misery may be given to others by their passion.

In a letter from a wife whose husband left her for a younger girl there is this sentence: "I loved my husband dearly, but it didn't make any difference to the girl—she broke up my family three months ago."

The letter continues, "He tried to come back to me and sent her away, but she would not leave him alone. She kept telephoning him and worrying him, until at last he went to her again. These girls don't know the heartache they bring—but I wish they would always try to see the man's wife before they steal a husband. Then they might understand how a women feels."

I think the only thing that can really put these triangular problems to rights is an influx of common sense and decency into the husband. He is primarily in the wrong—he has broken his vows to his wife, betrayed and encouraged a girl into loving him, and (in the above case) has endeavoured to retain the affections of both women without adequate return to either.

But love is blind. The wife is blind to her husband's weakness and irresponsibility, the girl is blind to the unhappiness of her future as the companion of a married man, the man is blind to everything but his own transient pleasure.

Supposing the girl did come to see the wife. To her, the wife would be nothing more than a woman who had legal rights over her lover—but no human rights. The wife would regard the girl as a usurper, an irresponsible flirt and a hard-hearted hussy. Could such a meeting and such points of view help in any way?

I am not laying down the law on the matter, however. In some cases, where the women concerned are more dispassionate, a meeting between wife and the other woman might be very useful.

AWKWARD AND DANGEROUS

We are very great friends as a family, with another family where there is a boy of my own age. This boy and I grew up together and have always been good pals until a few weeks ago when we went out together and were walking home.

He suddenly began to make violent love to me, dragged me off the road and I was terrified what would happen. Fortunately some people came along and I was able to escape, but since then I have been terrified of him.

I never want to see him again, and never would but for the fact that his parents and my parents are great friends. I don't like to tell anyone about it.

✖ Tell your own parents about it, at once, my dear. If they are friendly with the other family, it need make no difference—but they can see that you need never be alone with this young man again.

If necessary, they could tell his parents something of the matter, but the prime necessity is to protect you in future.

UNUSUAL AFFECTION

A fairly senior woman member of our office staff has become far too friendly with another girl—a junior—and seems to be absolutely devoted to her. The rest of us feel that this is unfair to the junior, as well as being abnormal.

Is there any way in which it might be stopped? It is upsetting us all.

✖ It is quite possible, you know, that there is nothing abnormal in this. A lonely middle-aged woman can strike up a half-motherly, quite harmless, friendship with a younger girl.

If, though, abnormality should enter into the case, there is little any outsider could do. The girl would in all probability find other friends and the tragedy would not be hers, but the older woman's.

As neither of these people has asked for help or advice, I can really only suggest that no one interferes with what is essentially a private, personal matter.

Virginia Ironside

Don't just sit and worry   share your problem with Virginia who's always here to help

Bitter news

I'm a single parent with a three-year-old son and his father comes to see him each weekend. The problem is that I still love him and can't stand the thought of his being with other women. I can't stand, either, the poverty and loneliness he has left me with. The resentment towards my ex-boyfriend has got so bad that I've started saying things to my son to turn him against his father. We also argue bitterly in front of him and I'm afraid I am getting confused. My own parents did with me. I was very unhappy so I know how awful it is but I can't stop myself, I wouldn't go to counselling or Gingerbread by the way, so please don't suggest this.

I'm sorry you're so dead set against seeking help—but I understand that it is hard to think positively when you're feeling so low. Still, you owe it to your young son to sort out this problem. Couldn't you kill two birds with one stone by asking your ex-boyfriend to take your son out for the day and have him for a night at weekends? This way you wouldn't argue in front of him, and you have some time and an evening to yourself. This is where Gingerbread would help, I have to say—you'd meet others so you could pool your resources and get out. Don't run your son's father down to him, whatever you do. If you do, apologise later and say that you didn't mean it. Small children can understand this sort of contradiction up to a point—but of course it's better not to do it in the first place. I'll send you my leaflet on single parents which has lots of addresses and sources of help.

Did he rape me?

I stopped going out with my boyfriend six months ago but last week he came round drunk and raped me. I went to report him to the police but my friend says that as I allowed him to have intercourse with me in the past, then this couldn't be classed as rape. Surely this isn't right?

No, indeed it is not and you should report him. I have to say, though, that as you have left it a while and you haven't got medical evidence, which usually needs to be seen as soon as possible after a rape takes place, you may find it hard to prove. Added to this, your boyfriend will obviously say that you were willing and, because you used to sleep together, you will find it harder to convince people he raped you this time. You will find support, by the way, from your local Rape Crisis Centre. The central number is 01-837 1600. While I'm on the subject, I'd like to correct what I said in February about a woman who consented to intercourse but withdrew her consent *during* intercourse. I said the man would not be found guilty of rape but in fact a New Zealand case has been drawn to my attention which holds that he would be found guilty. Whether one thinks this is right or not, this is the law which would be followed in this country.

Sexual fears

Until last year when I was 22 I was a virgin—simply because I was terrified of my 'first time'. I even thought I was gay—and had an affair with a close girlfriend which ended and left me unhappy. But I now realise what's wrong with me. I have a fear of someone entering me. Even with my girlfriend there were moments when I got very tense. I find myself turned on to men now but only if there's no risk of anything happening. Can you help me?

I certainly think you need to talk this problem over with a sex counsellor. The Marriage Guidance Council has psychosexual clinics and it's there you'd probably meet a trained counsellor who would be able to help you with your fears. You've got so far working things out already, you are intelligent and highly-motivated. This problem is not at all uncommon. I'll send you my sex leaflet which has other sources of help but the Marriage Guidance (under M in the phone book) would be the best, I feel.

ask me anything

Help when you need it... our agony aunt Sue Frost answers your letters

THE SPARKLE HAS GONE

The millennium really started me thinking about my life. I've been married for 12 years, but I just don't love my husband the way I used to. We have two lovely kids, but we don't seem to have fun together any more. We never go anywhere, never do anything exciting. Even New Year's Eve was nothing special, just the same old routine. I've got a couple of friends whose marriages ended and they're always out having a good time. I know it's hard when you've got kids, but I long to be single again. Surely everyone has the right to grab their chance of happiness?

Don't let that millennium hype get to you. It's a great chance for a new beginning, but that doesn't mean you should ditch everything that's gone before. Like many marriages, yours has gone a bit stale. You've lost touch with each other, you've let routine dull the love you once knew. But you can wake up your love life with just a little effort—talking about the things that matter, setting aside time to be together, planning a new future. Happiness may be closer to home than you think, and your single friends may secretly envy you.

He's perfect—but he's so young

At 38, I've finally met a man who's right for me. Trouble is, he's 10 years younger. Friends say it's no one else's business, but my family have made jokes about 'cradle-snatching' and I'm wondering if it will work out. He says he loves me and wants us to live together, even try for a baby before I'm 40. I know he's the one I've been waiting for, but I never thought it would be so complicated.

What's so complicated? Your man loves you and he wants you to have his baby. You've finally met the man you've been waiting for. It sounds just fine to me, and the only cloud on the horizon is your oversensitivity to family jokes. Tell them that you're the one who's laughing because happiness has come your way at last. Then stop worrying about it and be happy. You're looking for problems where no problems actually exist.

He sees his ex every week

My partner spends Sundays with his children and his ex-wife. They were divorced before we even met (she left him for another man, who's since left her) and I know she very much regrets what happened. My partner insists he loves me now, and when the time is right he'll bring the kids to our place. But I wonder about the time they spend together. What if his ex tells him she wants him back?

Your partner's ex may well want him back, but that doesn't mean he wants her. Ask how soon you can meet the children, and tell him you understand his need to spend time with them. Showing sympathy while keeping your natural worries at bay will build a solid base for your relationship. Encourage him to talk about his marriage—that way, he'll know he can trust you with his feelings.

My lover is free, but do I want him?

For two years I've been having an affair with a married man. He seemed like my ideal lover—kind and reliable, the sort of man I'd want to marry, but that seemed unlikely as he was committed to his wife. Last week, however, he confessed to our relationship and his wife threw him out. Now, suddenly, I'm not sure about us any more. Perhaps we were better as lovers than live-in partners?

Maybe your ideal man isn't so kind and reliable after all. Maybe he's just a cheat whose wife is glad to see the back of him. Maybe his 'commitment' to his marriage was just an excuse to fob you off, and he's only free now because his wife finally decided she'd had enough. But whatever the truth, your doubts ought to make you think again, not only about him but about your future relationships too. Deception is a poor beginning for anything worthwhile.

I can't face life without him

I'm 19, I've been with my boyfriend three years and I've never wanted anyone else. I thought he felt the same way, but now he says we're too young to be tied down. He says that we need some time apart to see how we both feel, but I know this really means it's over. He says he still cares for me, and he rings me at weekends to check that I'm OK. But I'm not OK, and I never will be until I get him back again.

There's no way to get a man back if he's determined to be free. And your boyfriend has been honest with you. He's not ready for commitment and he's tried very hard to make the break as gentle as possible. So be brave and let him go. Lean on your family, get out with your friends and be kind to yourself. It seems impossible now but, believe me, you will recover.

24-HOUR HELPLINE

Just dial the number below at any time, day or night. Calls cost 60p per minute and last approximately three minutes each. We hope you find these recorded advice lines useful

I'M ALWAYS ANXIOUS 0906 616 7946	I FEEL SO LONELY 0906 616 7949	MY PARTNER HITS ME 0906 616 7952	SHOULD I GET DIVORCED? 0906 616 7955
WHY AM I SO DEPRESSED? 0906 616 7947	I CAN'T HAVE AN ORGASM 0906 616 7950	I'M PREGNANT— IT WASN'T PLANNED 0906 616 7953	WE'RE SPLITTING UP 0906 616 7956
I'M SO JEALOUS 0906 616 7948	I DON'T HAVE ANY CONFIDENCE 0906 616 7951	HOW CAN I IMPROVE MY SEX LIFE? 0906 616 7954	HE'S HAVING AN AFFAIR 0906 616 7957

THERE IS
A WAY—

* * *

Perhaps You Are Anxious About A Little Social Matter, Perhaps You Need Advice In The Choosing Of A Career; We Will Gladly Advise You. If You Would Like Our Help, Address Your Letter To: THE EDITRESS, c/o WOMAN AND HOME, The Fleetway House, Farringdon Street, London, E.C.4, Enclosing A Stamped, Self-Addressed Envelope For A Personal Reply.

I have just received what I believe is known as a left-handed offer. In other words, a married man has suggested that he and I should have a love-affair. He has not said anything about getting a divorce and, as he is in a good social position, I am sure he would not want to break up his marriage.

The worst of it is I am strongly attracted to him, and he is just the one man I should have liked as my husband. I am twenty-four and he is thirty-six, and the days when we don't meet are quite blank for me.

He gives me lovely presents and we have such wonderfully happy times together that I just don't know how to do what I know I should do, and that is to send him out of my life.

I DO not believe for one moment that you would filch anything valuable belonging to another. You wouldn't do anything like that because, for one thing, it would be dishonest. Nor would you read letters intended for someone else because that would be mean. And least of all, would you deliberately set out to cause distress—let alone misery—to a fellow human-being. None of these things is permissible, according to your code of behaviour.

But is there one great exception to all these rules of yours? Something which makes this code break down entirely—when you happen to fall in love with a married man?

Reactions to this very serious matter are sometimes not at all what they should be. From the very first, the wife doesn't seem to come into some women's thoughts about it.

Do you never say to yourself, "If I don't stop myself loving this man—and stop it here and now—there's going to be bad trouble for another woman and perhaps a clouded future for her children, as well as herself."

I am sure you are fundamentally a nice girl and it often makes you quite unhappy to think what a muddle you're in. But you won't give it up. You can't. My dear, a man who lets one woman down will let another down, because it comes from a flaw in his own nature, more than from circumstances. That very weakness which made him transfer his affection from the woman he married to the girl he fell in love with, is quite likely to betray him again . . . perhaps when you are no longer a novelty. If this were to happen, you would understand just how his wife feels.

So, I want you to draw back, here and now, with all the strength of will you possess.

Never mind if it hurts. Take the long-term view, and resolve never to be the thief who steals another woman's most precious possessions . . . her home life, her security and her husband.

★ ★ ★ ★

NOT TOO OFTEN

How many evenings a week do you think a girl should spend in her own home? I am Secretary of a Youth Club, attend evening classes, and belong to a Debating Society, and I am also taking tap-dancing lessons. And now my parents complain that they never see me.

Is this fair? I am an only child and my age is eighteen.

IT is good for a young girl to have as many constructive activities as you favour, my dear. But don't forget the intense pleasure your parents get from just seeing someone young and gay around to talk and laugh and brighten things up for them. Don't grudge them two evenings a week, even if it means re-arranging your programme.

And a word in your ear: the girl who before marriage gets the habit of thinking a quiet evening spent at home intolerably tedious will have formed a habit which may well undermine the happiness of her own home, one day. So do take my advice for your own sake, too.

INTANGIBLE GIFTS

Mine is rather an unusual problem. The girl I am in love with is an orphan, and has quite a good income of her own, left to her by her parents who were both well-off.

I am almost sure she loves me but I just can't bring myself to ask her to marry me, as I earn far less money than she has and I am afraid of people thinking me a fortune-hunter.

Could any marriage possibly be happy under these conditions?

I HAVE known several supremely happy marriages in which the wife happened to have more money than the husband. And it would seem hard that the possession of a personal income should deprive a girl of happiness with the man she loves.

Money is a tangible thing, whose value can easily be computed, but there are other, intangible gifts you would bring her.

Marriage, a home, children, love, and companionship. What is the value of such things to a woman? Ask her, and see what she says; and the best of luck to you both.

ACTIVITY

> If you have been working in groups, share your ideas with the whole group. Whatever your chosen working method consider the following:
>
> - Do you think the language of the two texts on pages 136 and 139 suggests that the two agony aunts have different personae? Give some examples of language as evidence for your opinion.
>
> - Do you think the female audience for the two texts are assumed to have different concerns and attitudes? Again, give some linguistic proof for your ideas.

RESEARCH
PATHWAYS

> Articles from different magazines aimed at the same interest group, e.g. gardening, interior design, music; the same feature in magazines aimed at different groups, e.g. horoscopes, fashion, in men's and women's magazines.

Websites

Just as there are online newspapers, so there are e-magazines, some of which are online versions of their paper equivalents, while others have been started from scratch as new 'e-zines'.

To find an online equivalent of a paper magazine, all you normally have to do is to type in the magazine name, followed by .com or .co.uk, for example:

http://www.menshealth.co.uk
http://www.elle.com

but there are new e-zines such as:

http://www.charlottestreet.com
http://www.handbag.com

You might be interested in contrasting magazines which appear to be aimed mainly at a white audience with some that appear to target other communities, in which case go to:

http://www.blink.org.uk
http://www.blackliving.com
http://www.ebony.com

Comics

Despite the attractions of television, video and computer – or, perhaps because of these media – the comic is still as popular as ever. Nowadays, comics are more and more closely tied to their moving image counterparts from a marketing point of view: Batman, Star Trek, and Superman all have both film and written comic outcomes where each medium supports the other commercially; a renewed interest in fantasy films – such as the 'Terminator' and 'Alien' series – has generated new larger-than-life gladiators who are female as often as male, and has brought the conventions of the comic format to the big screen; and new comics have sprung up in the wake of the success of computer games, such as *Playstation*. Even so, old favourites such as *The Beano* and *Bunty* still survive. What are comics all about? Are they just ephemera, unworthy of serious thought and analysis, or can we learn interesting things about society by looking at their language? This section will help you to start exploring these questions.

ACTIVITY

Make a list of the comics you read as a child.

Do you remember particular characters in the comics you read? Why were they memorable – what did you enjoy about them?

What purpose do you think comics served, in your childhood?

Is there such a thing as 'the language of the comic'?

See if you can come up with a description of the comic format. Think about:

- the titles of comic strips

- aspects of layout

- the structure of the narrative

- typical vocabulary which might be used by the characters

- presence or absence of a narrator, and, if present, what the role of that character might be

- the whole purpose of the comic story.

ACTIVITY

> If you were working in groups, share your recollections of your childhood comics with others in the whole group.
>
> What particular types of comic were popular with you all?
>
> Do male and female group members find that they read different sorts of comic, and if so, why do you think that is?
>
> Do you agree on the purpose of the comic?
>
> If you were working alone, ask six informants about the comics they read as children, and about their views on the purposes of comics, then try to summarise your findings from this research.
>
> Whatever your chosen working method, consider the following:
>
> • How far were you able to define the comic genre by its typical language use?

ACTIVITY

> Just as an interesting perspective can be gained in analysing changes in the language of newspaper and magazine genres over the years, so comics can be scrutinised in the same way.
>
> Early comics were targeted at boys, but proved so popular that girls' comics soon came onto the market.
>
> Look at the four extracts on pages 144–148.
>
> They are all openings from stories in *The School Friend* comic. *Extracts A, B* and *C* are from the 1957 annual; *Extract D* is from 1959.
>
> Explore the language of the comics, particularly the following:
>
> • The representation of the non-English characters, including the language they themselves are given
>
> • The language and roles given to the English characters
>
> • Representation of gender
>
> • The language used by the 'narrator'
>
> • How far the comics follow the outline you brainstormed in the previous activity
>
> *continued*

- Which of the stories do you think are derivative of the types of stories that would have featured in boys' comics? Are there any details in girls' versions that would not have appeared in stories aimed at boys?

- What do you think were the functions of the stories, and what effect would they have had on their working-class readership?

ACTIVITY

If you were working in groups, share your responses to the 1950s comics.

Whatever your chosen working method, write a summary on the following:

- How do the comic stories construct ideas about gender, race and class by their language use?

ACTIVITY

Now look at a contemporary comic narrative – 'The Legend of Little Plum', from *The Beano*, 1998, on page 148, and consider the following:

- What aspects of the 'Little Plum' story are familiar in comic narratives? What overall purpose does this type of plot have for young readers?

- Would you say this story presented a stereotyped picture of American Indian culture? Are there any differences between this story and the earlier texts in terms of stereotyping?

- How would you describe the language used by the characters in the story?

- Examine the various aspects of verbal language in the cartoon strip. What else is the language being used for, apart from the characters' speech? (For example, who is the narrator, and what is the function of the various 'labels' in the pictures?)

EXTRACT A

EXTRACT B

EXTRACT C

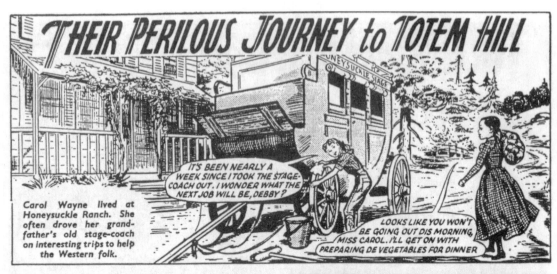

EXTRACT D

That's all from our chum Plum for now. Don't miss NEXT WEEK'S DOUBLE-SIZED, BUMPER BEANO BIRTHDAY ISSUE!

RESEARCH
PATHWAYS

> Comics aimed at different genders; older comics compared with modern equivalents, e.g. *Batman*; representations of race in comics; a comic compared with its moving image equivalent; parodies of comic style compared with their originals, e.g. *Viz*; the use of comic formats in other genres, e.g. advertising.
>
> Do some research among young readers on their reading habits in relation to comics. For example, do boys and girls prefer different comics, and why?
>
> What are the children's ideas about what makes a comic story successful and enjoyable?
>
> Are children aware of stereotyping in the stories they read?

Advertising

It will come as no surprise that there is no such thing as the language of advertising, although there is a variety of linguistic techniques that advertisers use to achieve their common purpose: to persuade. While the features of language use in advertising may vary, then, at least the overall purpose remains constant.

While it is perfectly possible to analyse the language of a single advertisement in order to see how the persuasive function is built up by its use of language, some of the more interesting researches involve comparing and contrasting. As with other areas of language research, a comparison can enable the researcher to see very clearly, and to point out to readers, why certain choices of language have been made. This is particularly true where the products being advertised in the different texts are the same or similar.

One difficulty for the researcher is that, even if a comparison is the chosen route, there are so many advertisements to choose from: on what basis can any comparison be made?

ACTIVITY

> Think about some of the ways it would be possible to make a comparison of advertising texts for the same or similar products. Fill out the list below by giving more examples of advertised products in each case.
>
> Comparison according to:
>
> • age of target groups: holidays
>
> • gender of target group: cars
>
> • social class of target group: alcohol
>
> • ethnic origin of target group: hair products.

continued

When you have exhausted your list above, move on to think about the ways that the various groups you have described above feature in adverts, rather than being the target audience.

The way in which a particular group of people is used to sell products could be the constant factor that links your advertising texts together. For example, what products use images of children to sell to adults; images of women to sell to men; images of middle-class people to sell to working-class people; images of black people to sell to a white audience, or vice versa?

ACTIVITY

Look again at the comparisons that it might be useful to draw in a piece of language research on advertising.

Can you now see a number of ways in which you could achieve an interesting focus in researching this area?

ACTIVITY

Look at the five adverts on pages 152–156, which are all selling Ovaltine.

Adverts A, B and *C* are from the 1930s; *Adverts D* and *E* are from 1988.

Read all the adverts through carefully before you start your work, then try to decide:

1. Who is speaking to the reader in each of the adverts – what kind of person do you imagine behind the words? What kind of persona has been constructed by the advertisers, to speak to you?

2. Looking carefully at the language, try to work out how you decided the answer to question 1. What was there in the language to make you come to certain conclusions? Try breaking the language down into the five language levels:

Graphological level: Images, typeface, layout, etc. and their effects

Semantic level: Words and phrases – their connotations, their level of formality or intimacy, and the reasons for this

Grammatical level: Particular structures used, e.g. commands, statements, structures that imitate speech

Phonological level: Use of sound patterning

Discourse: Whether the whole texts are working in a particular way, e.g. imitating a specific genre

ACTIVITY

If you have been working in groups, share your ideas on how the Ovaltine adverts work. Whatever your chosen working method, write a summary on the following questions:

- What are the different techniques being used, in order to construct a particular persuasive voice to speak to the readers of the Ovaltine texts?

- When Ovaltine was first marketed, it was sold through chemist's shops, as a medicinal product. Is there any evidence of this in the older adverts?

- From your analysis of the data, are there any other changes evident in the way the product has been marketed over the years?

RESEARCH PATHWAYS

Go back to the ideas you generated during the first activity you did in this section on advertising. Can you see, from those ideas, and from the other analytical work you have done in this section, how you could explore advertising with a detailed focus on language use?

Here are some more ideas: adverts which sell 'delicate' products, e.g. sanitary towels, condoms; TV adverts featuring particular groups, selling to particular groups or using certain linguistic techniques; use of foreign languages in English adverts; techniques used on advertising hoardings; comparison between English and American adverts for similar products; techniques used in 'junk mail' adverts.

Websites

There are so many Internet websites devoted to advertising that it would be impossible to list them all here. If you want to know whether a company advertises itself and its products, try typing in its name, followed by .com or .co.uk. Here are some examples:

cocacola.com
volvo.com
ikea.com
kelloggs.co.uk
tesco.co.uk

There are also sites that are about advertising itself. For example, if you go to:

http://www.adbusters.org

and click on 'spoof ads' or 'uncommercials', you will find material that you might call 'alternative advertising'.

ADVERT A

"I look forward to 'Ovaltine'" says Mary Glynne

"I HAVE always looked forward to my cup of 'Ovaltine' at night after working at the Theatre," writes Mary Glynne, the famous actress. "My kiddies love it too . . . . and I always keep a good supply for them."

As a "night-cap" for ensuring sound, refreshing sleep and as the daily beverage for giving glorious health and abundant vitality, there is nothing to equal delicious "Ovaltine."

This complete and perfect tonic food is 100 per cent. health-giving and energy-creating nourishment. It is scientifically prepared from the highest qualities of malt extract, fresh creamy milk and new-laid eggs. Unlike imitations, "Ovaltine" does not contain any household sugar to give it bulk and to reduce the cost. Nor does it contain a large percentage of cocoa. Reject substitutes.

'OVALTINE'
Tonic Food Beverage

Prices in Gt. Britain and Northern Ireland,
1/1, 1/10 and 3/3 per tin.

P. 943

ADVERT B

An open letter to Mothers of fast-growing children

THOSE children of yours are growing so rapidly. The great concern of every mother must be that the growth shall be normal and regular, and that body, mind and muscle shall develop at the same rate.

Many children show a tendency to outgrow their strength. They become listless and disinclined for play. Their appetites are capricious and they are often weak and ailing.

Healthy and normal development depends almost entirely on correct diet and proper nourishment. Every particle of the material used in creating energy and building up the brain and body is obtained from food.

Growing children need more nourishment than ordinary food supplies. That is why "Ovaltine" should be their daily beverage. This delicious food-drink supplies, in a concentrated, correctly balanced and easily digested form, all the nourishing elements and vitamins that are essential for healthy growth.

"Ovaltine" is prepared from creamy milk, malt extract, and eggs from our own and selected farms. These are Nature's best foods. Eggs supply organic phosphorus—an essential element for building up brain and nerves.

The addition of "Ovaltine" removes the objection many children have to plain milk. "Ovaltine" renders milk more digestible, and therefore more beneficial. The nourishing value of all ordinary foods is increased when "Ovaltine" is the daily beverage.

Give your children "Ovaltine" instead of tea, coffee, etc. They will grow up strong and healthy—with sturdy bodies, sound nerves and alert minds.

"Ovaltine"

"OVALTINE" BUILDS UP BRAIN, NERVE AND BODY

Prices in Gt. Britain and Northern Ireland, 1/3, 2/- and 3/9 per tin.

P677

ADVERT C

Specially drawn by Fortunino Matania, R.I.

Great Builders of History
The Builder of the Taj Mahal

BY common consent the most beautiful building in the world is the Taj Mahal —the memorial of an undying love and the mark of the genius of its builder, the Emperor Shah Jehan. The passing centuries have not impaired its white gleaming marble, and it stands to-day in all its pristine loveliness.

Through centuries the Great Builders have permanently enriched the world with work of beauty and strength. Their work has endured, whereas even empires have often decayed and perished.

And the builders of health have also done work of enduring merit. Their mission has been to upbuild the health of humanity and increase the sum of human happiness. Among such health-builders 'Ovaltine' deserves and holds an honoured place.

This supreme tonic food beverage is recognised all over the world as the best natural means of giving and maintaining good health.

Prepared from malt extract, fresh creamy milk and new-laid eggs, it owes its supremacy to the quality and proportions of these ingredients as well as the scientific process of manufacture.

Delicious to taste . . . easy to digest, 'Ovaltine' is also a complete and perfect food beverage. It supplies, in unequalled abundance, all the protective vitamins and essential food elements that restore vitality and build up perfect health of body, brain and nerves.

Remember—for health and energy all day and peaceful sleep all night, 'Ovaltine' is supreme. It stands in a class alone—there is nothing like it.

'OVALTINE'
The Supreme Builder of Health

Prices in Gt. Britain and N. Ireland,
1/1, 1/10 and 3/3 per tin.

Novartis is the copyright holder and trade mark owner of Ovaltine™

ADVERT D

THE THAIS DRINK OVALTINE FOR ENERGY.

WHAT MUST THEY THINK OF US DRINKING IT BEFORE BED?

Before indulging in any strenuous activity, the good citizens of Bangkok indulge in a glass of their favourite energy drink. Ovaltine.

In Thailand, you see, Ovaltine is seen differently than here. They actually call it the 'kick the day off' drink.

And, as Ovaltine contains the goodness of malt extract, barley and eggs, why not?

Malt extract, for example, is well known as an instant and long lasting source of energy.

Eggs are an excellent source of protein. And Ovaltine provides calcium in abundance. (Through both the milk powder it contains and the fresh milk you make it with.)

And, a mug of Ovaltine contains no artificial flavour, no added sugar, colour or preservatives.

Small wonder then that your average Thai football team insists on a pre-match mug.

Yet, despite the reasoning, it still sounds somewhat strange. Then again, if we think they're odd, what must they think we are?

Bonkers, probably.

Novartis is the copyright holder and trade mark owner of Ovaltine[tm]

ADVERT E

THE FRENCH CLAIM OVALTINE RESTORES ENERGY. IS THAT WHY THEY DRINK IT FOR BREAKFAST?

After a très fatiguant night on the town, what have this French couple headed straight for?

Hair of the dog, perhaps? Or black coffee? No . . . a cup of Ovaltine. Yet with the old joie de vivre obviously exhausted, why do they drink something to help them sleep?

Thing is, they don't. They drink it to wake up. Because your Continentals believe that the malt extract, barley and eggs in Ovaltine are a real boost to a flagging body.

Malt extract for instance, is well known as an instant and long lasting source of energy.

Eggs are an excellent source of protein. And Ovaltine provides calcium in abundance, through both the milk powder already in it and the fresh milk it's made with.

What's more, thanks to a generous helping of cocoa, Ovaltine has a delicious chocolatey taste.

Yet Ovaltine contains no added sugar, no artificial flavour, colour or preservatives.

So the moral of the story is this, girls. If a Frenchman offers you a mug of Ovaltine, beware.

He actually means next morning.

———— ISN'T IT TIME YOU WOKE UP TO OVALTINE? ————

Written genres

As an alternative to looking at language use within a particular written genre such as newspapers or comics, it is also possible to compare whole genres of writing with each other.

The aim for this type of study is to highlight the varying ways in which different types of writing work. Obviously, trying to characterise whole genres of writing by looking at many texts is an impossible task in a small study which requires detailed analysis, but small extracts of different types of writing can be compared usefully if the writing has some subject matter in common.

As an example of how this could work in practice, try the activity below.

ACTIVITY

Look at the two texts on pages 158–159.

Both texts are about zoos, and are aimed at primary age children. The text in *Extract A* is from a topic book called *Zoos* by Miriam Moss (Wayland (Publishers) Ltd, 1987), and would fall into the non-fiction category; the text in *Extract B* is from a story book entitled *Who's For the Zoo?* by Jean Ure (Orchard Books, 1989).

Read the two texts through carefully, and consider the following:

How does the language differ between the texts? Explore a range of different language levels. Here are some examples of the ideas you could consider for each level:

Graphological level: How do the layouts of the two texts compare? How are pictures used, and do their contents relate to the texts in the same way? How are different groups of people represented by the images in the texts? What are the typeface conventions of the texts?

Phonological level: Is sound patterning used in either of the texts, and, if so, why?

Semantic level: Do the texts differ in the amount of information-processing they require – in particular, the amount of new nouns introduced? How are terms which refer to people and things linked together throughout the texts? How are different groups represented by the language used?

Grammatical level: Are particular structures used frequently in either of the texts? How complex are the structures used: for example, are subjects and verbs close together, to aid comprehension?; are active verbs preferred to passives?

Discourse: What is the overall shape of the texts: for example, do they

continued

have a thematic, or a narrative, structure? What purposes are served by the different texts, and how is this reflected in the language used? How would the books from which these extracts were taken differ in their overall organisation?

ACTIVITY

If you have been working in groups, feed back your results to the whole group.

Whatever your chosen working method, write answers to the following questions as fully as possible for your file:

- What general conclusions can you draw on the differences between the two zoo texts in their purposes?

- How do any differences in purpose show up in the language used?

EXTRACT A

Taking Care of the Animals

The people who look after the animals each day are called zoo keepers. The keepers start work early in the morning. They change into their overalls and wellington boots, pick up a stiff broom, a bucket, a hose pipe, some **disinfectant** and a shovel and go to the animal **enclosures** to clean them out. Then they prepare the animals' meals for the day in freshly scrubbed steel dishes. The keepers give a list of all the food they might need for the week to the senior keeper who orders it from the zoo's food store.

It is important that the animals are given the right kind of **diet**. Some animals need special food to keep them healthy. For example, a vampire bat needs 0.25 litres of fresh blood a day and the apes like blackcurrant juice. The reindeer need a **lichen** that is imported from Iceland.

Penguins are fed by hand, by their keeper, as they will not pick the fish up from the ground. Each type of animal is given a carefully planned diet to suit its needs.

EXTRACT B

Chapter 3

Next day, Miss Lilly said that in preparation for their visit to the zoo they were all going to write animal poems.

"I want you each to write a poem about the animal that you've put in your picture. Before you start, I want you to sit and think for a bit about the animal that you've chosen. Think what sort of animal it is ... what sort of personality it has. Whether it's a comedian, like a monkey, or fierce, like a tiger. Try to make the poem be a bit like the animal."

Cameron, who didn't like writing poetry, said, "How can a poem be like a yak?"

"Well, now," said Miss Lilly, "it seems to me that that's up to you ... you're the one who likes yaks! But I'll tell you what, it's a jolly good animal for rhymes! Yak, back, clack ... there are dozens of them!"

"Think how you'd feel," said Bader, "if you'd chosen gorilla."

"Yes, or hippopotamus," said Alison.

Miss Lilly pointed out that poems didn't *have* to rhyme. "We mustn't get hung up about it."

"But nothing rhymes with hippopotamus!" wailed Alison.

Cameron, suddenly inspired, said, "That's your problem!" and snatched up his pen. Soon everyone was writing hard. Miss Lilly said that tomorrow she would read out some of the best ones.

In the classroom at break they groaned at each other about the animals they had chosen.

"Hippopotamus is *aw*ful. There isn't anything."

"You want to try *pen*guin."

"You want to try *pan*da."

Pavindra, always considerate of other people's feelings, said, "What about Catherine, with gorilla?"

They looked round for Catherine, but she wasn't there. Not even Soozie knew where she had gone.

RESEARCH PATHWAYS

Comparison of further texts from different genres where the content is the same or similar, e.g. holiday brochure compared with an extract from a geography textbook, both about Amsterdam; a piece of prose or poetry on the same subject.

Spoken genres

Just as different genres of writing can be compared, or a particular genre of writing analysed for what makes it distinctive as a variety, so spoken texts can be treated in the same way.

The section on writing concentrated on the comparative approach across different genres; this section will look at one particular genre of speech: the wedding ceremony.

ACTIVITY

Any genre, spoken or written, is likely to have a number of possible variations, and the genre of the wedding ceremony is no exception. Compile a list of possible variations by continuing the list of ceremonies below:

- Civil ceremony

- Quaker ceremony

When you have finished this list, read through the Civil (*Text A*) and Quaker (*Text B*) ceremonies, and consider the differences between them.

TEXT A

Wedding Ceremony
Preamble

We are gathered here on this happy occasion to celebrate the marriage of ABC to XYZ a ceremony to which all present are witnesses.

According to the law of this country marriages have to take place in registered buildings therefore I must inform you that this place in which we are now met has been duly sanctioned according to law for the celebration of marriages.

I am also required to ask of you that if any person present knows of any lawful impediment to this marriage he should declare it now.

Now A and X, before you are joined in matrimony I have to remind you of the solemn and binding character of the vows you are now each about to make. Marriage according to the law of this country is the union of one man with one woman voluntarily entered into for life to the exclusion of all others.

continued

Declaration

I declare that I know of no legal reason why I, ABC, may not be joined in marriage to XYZ.
I declare that I know of no legal reason why I, XYZ, may not be joined in marriage to ABC.

Contracting Words

I, ABC, take you, XYZ, to be my wedded wife.
I, XYZ, take you, ABC, to be my wedded husband.

Now A and X you have both made the declarations prescribed by law and have made a solemn and binding contract with each other in the presence of your family and friends (or witnesses) here assembled and have symbolised your marriage by the joining of hands and the giving and receiving of rings (a ring).

I have great pleasure in informing you that you are now husband and wife.

TEXT B

Wedding Service

Suggested form of words

Friends, I take this my friend .... to be my wife, promising through divine assistance to be unto her a loving and faithful husband so long as we both on earth shall live.
Friends, I take this my friend .... to be my loving husband, promising through divine assistance to be unto him a loving and faithful wife so long as we both on earth shall live.

The service takes place at an ordinary meeting. When the couple are ready to stand up they say these words to each other.

Discuss the differences (and similarities) between these and any other wedding ceremonies you are familiar with. If possible, collect a range of different ceremonies for further exploration. There are often differences within one type of ceremony that can also be explored. For example, in the civil ceremony you studied earlier, people are allowed to make up their own form of words at certain points in the ritual. The document from which the ceremony was copied, issued by Lancashire County Council, has some suggestions for vows during the exchange of rings, for example: 'I give you this ring as a symbol of my love and affection. Please wear it with a feeling of love and pride now and always.'

continued

Here are some examples of aspects of language you could focus on in looking at further texts:

- The purposes served by the language ritual we call a 'wedding ceremony'

- How far the different ceremonies have fixed structures and utterances from which no deviation is allowed, and why

- The various statements, promises, etc. covered in the different ceremonies, and what conclusions can be drawn from any differences

- Differences in the vocabulary used in the ceremonies

- Differences in grammatical structures used

- The overall effect or message given by each of the texts

ACTIVITY

If you have been working in groups, share your ideas on the differences between the wedding ceremonies.

Whatever your chosen method, write a summary for your file on the following question:

- What are the differences between the various ceremonies, and how are these differences reflected in the language used in each text?

RESEARCH PATHWAYS

Further studies of particular spoken genres: for example, nursery rhymes, playground songs and other spoken rituals used by children; jokes; anecdotes; sermons; speeches.

Answers

Activity page 122

(a) Autobiography – Jeanette Winterson, *Oranges Are Not the Only Fruit* (Pandora, 1986)
(b) Fairy Tale – 'Blubber Boy' in Angela Carter (ed.), *The Virago Book of Fairy Tales* (Virago, 1991)
(c) Detective Novel – Sara Paretsky, *Toxic Shock* (Penguin, 1988)

4 Comparative linguistics

Comparative linguistics involves comparing English with another language.

Data for exploration of this area is normally gathered from a speaker or writer whose first language was not English, and the purpose of the study is to compare the features and/or the functions of the two languages.

Features and functions

Features are the 'raw ingredients' of language – the various language levels which were explored in previous units: sounds, written symbols, vocabulary, grammatical structures.

English may differ from other languages on all these levels, and the aim of a project on the features of different languages will be to show how they vary in a systematic way.

Functions are the purposes which language serves. Speakers who have more than one language may use their languages for different purposes: for example, one language may be used in formal situations, while another is reserved for use in more familiar and intimate settings; speakers will also negotiate their language choices according to the confidence and competence of the participants.

A project on the functions of different languages used by a bilingual speaker will look for pattern of usage, and attempt to describe the factors which trigger the use of one language or another. Switching from one language to another is known as *Code Switching*.

Bilingual speakers also sometimes mix their languages together, using words or phrases from one language in another. This is known as *Code Mixing*. Certain words may be mixed for the same reasons as those that trigger code switching between whole languages.

Work on Code Mixing may also consider the way particular words from one language are integrated grammatically within another. This will involve looking at the shape and type of words mixed, and their position in sentences.

Code mixed words may, in addition, be explored for their semantic content: are the mixed words from the same or similar areas of the vocabulary system, and, if so, why?

The features of different languages

ACTIVITY

Below is an extract from a conversation between a mother and daughter from the Manchester area who both speak Afro-Caribbean patwa. In the extract, the mother is talking as she moves about the kitchen, addressing her teenage daughter, who remains silent. At the beginning of the extract, the mother has found some jewellery on the kitchen floor.

The top line is a phonemic transcription of the speaker's language; the bottom line is a written Standard English version of what she is saying.

Study the speaker's language and make notes on:

- her accent – compared with RP

- her vocabulary – compared with Standard English

- her grammatical structures – compared with Standard English.

hu	dɪs	am/mi	nəʊ	nəʊ	aʊ	dis	draːp/əʊt	ə
Whose is	this	um/I	don't	know	how	this	dropped/out	of

dɪ	bɒks	jʊ	nəʊ/wen	mi	wəz	pʊtɪn	ɪn	də/səʊ	am	hæv
the	box	you	know/when	I	was	putting	in	the/so	um	have

tʊ	pʊt	ɪt	bæk	dɛə/bəfɔ	mi	get	lʊs	ɒf	dem/fɔ	jʊ	æfi	gəʊ
to	put	it	back	there/before	I	lose		them/so		you have		to

mek	kjake	fi	mi/enitɪn	dalɪn/aɪm	næt	fʊsi/mi	nəʊ
make a	cake	for	me/anything	darling/I'm	not	fussy/I'm	not

pʊt	nəʊ badi	ɪn	ə det	mɪ	dɪər/enitɪn	jʊ	kən	əfɔd
putting	anybody	in	debt	my	dear/anything	you	can	afford

ACTIVITY

If you have been working in groups, share your notes on the language of the patwa speaker.

Whatever your chosen working method, present your notes in the form of an economic summary of the main patterns of language use you have found, by answering the following questions:

- What patterns of pronunciation difference from RP did you note?

- What grammatical features did you note as being characteristic of patwa?

- Were there any words where you had some difficulty categorising the variation: for example, whether you should call the variation an accent or a vocabulary feature? List these, and explain why categorising them was difficult.

ACTIVITY

Read the letters below, which are written by a French student learning English as a foriegn language at school to her English penfriend, Joann.

When you have read them thoroughly, make notes on the following:

- What aspects of English semantics (i.e. rules for choice of words and phrases) are causing the writer some difficulty?

- What structural (i.e. grammatical) aspects of English are a problem to the writer?

- Are there any aspects of English spelling or punctuation that are causing problems?

- For all the above areas, try to find patterns of usage, where the same feature occurs a number of times. Can you suggest how the writer's first language might be influencing her use of English?

67 Rue Delacroix
69002 Lyon

Dear Joann,
Thank you for your letter. I've just received it.
I hope you and your family are well. Here,
everybody is nice. The life is normal again: I go
to school, I do my homework (which is not
always funny) and I gave up eating too much...
though my mother is on a diet, she likes the
sweets you offer us and so am I.
I think you and your family are very nice to
me. Now I'm going to write you something
secret. I've been falling in love with a boy
handsome for two weeks. The more I see him,
the more I love him... And what about you?
I'm sure he likes you aswell. Well, I'm going
to write in French because I'm fed up with
writing in English (it's very dificult) and
it's a good practise for you...

67 Rue Delacroix
69002 Lyon

Hi Joann!
I think for you from the mountain where I'm spending my holidays very sunny and sportives. I hope that you and your family are fine as well. I'm happy, because March is approaching and I'll introduce you my family and my friends who would like to meet you as I talk about you.
See you soon.
Love Nathalie
x x x

Westem
croft
ns Lane
in
star
OAB

67 Rue Delacroix
69002 Lyon

Hi!
I've just received your two post cards, it's nice to you. I wish I had come to visit you and your family but when I heard of the date of the journey, it was too late. I had already plannified my holidays with my father otherwise, I would have come with great pleasure. I'm really disapointed not to be able to come (de ne pas venir te voir), because I keep a nice remembrance of you, your family and of the stay. It may be for a next time. As for you, if you want to come in France, there is no problems, you'll be alaways welcome. Well thank you again. I'm looking forward to hearing from you. See you one day.
Love Nathalie xxx

ACTIVITY

The functions of language

ACTIVITY

Below is some spoken data collected by A, a speaker of Punjabi, Urdu and English. The first dialogue is between this speaker and her sister, B, who has the same repertoire of languages. The second dialogue is between A and her mother, C, who is not yet fluent in English. Read the conversations thoroughly.

DIALOGUE 1

B: what's the tape running for

A: I've got to do something/for my English lesson

B: what/taping people

A: mm/I'm going to tape mum and dad

B: but no one will be able to understand it

A: I know/that's the whole idea/I'm not daft/you know

B: what do you want people speaking in Pakistani for

A: it's Punjabi actually/not Pakistani

B: alright then/Punjabi

A: it's for my language project/alright

DIALOGUE 2

The non-English words in the following are spoken in Punjabi. They have been transcribed using the phonemic alphabet. Underneath each utterance are two translations: first, a word-for-word translation, then a translation using English word order and idiom.

C: dʒa video tʃʊkæn

    go video bring

    go and get the video

continued

A: film   arndiheh

   film bought is

   have you bought a film

C: ɑh

   yes

A: which one/where is it

C: television   ne pitʃe

   television from behind

   behind the television

C: door   bænd kar sardhi heh

   door close it cold is

   close the door/it's cold

Now answer the following questions:

- Why does speaker A use English in *Dialogue 1*, but switch between languages in *Dialogue 2*?

- Look at the way English and Punjabi have been mixed in *Dialogue 2*. Is it possible to group the English words together, either grammatically (i.e. by their function) or semantically (i.e. by their meanings)?

- In Punjabi, verbs go at the end of an utterance. What effect has that had on the position of the English words used?

- Why do you think speaker B is surprised that her sister should be taping their family conversation?

- What do you think monolingual speakers can learn from looking at the way bilingual speakers use their languages?

ACTIVITY

If you have been working in a group situation, pool your answers to the previous questions.

Whatever your chosen method, do the following:

- Compile a list of questions you would like to ask bilingual speakers about their language experience, and try to find some informants to interview.

- If possible, interview a speaker of Punjabi and/or Urdu, invite them to talk about their own code switching and mixing, and ask them to respond to your analysis of the Punjabi/English data.

ACTIVITY

The researcher who taped the Punjabi/English data you have just been studying also collected some words of English origin from the speech of monolingual Punjabi speakers in Pakistan, during a holiday she spent there.

The words appear below, along with some details of how they were pronounced by the speakers, where the pronunciation seemed unusual.

The speakers all insisted that the words were Punjabi words, not English ones.

Consider:

* whether the words the researcher found could be grouped into particular semantic areas
* how the words came to be borrowed from English
* why the speakers did not recognise the words as being of English origin.

yes		no	nɔ
1–0		please	
taste	tɛəst	thank you	tɛənk
better	betər	hello	helɔ
sit down		stand up	
tape	tep	cassette	
radio	rediɔ	television	
book		copy	
pencil	pensil	pen	
chips	tʃɪpez	apple	epəl
biscuit	bɪskʌt	sweets	
police	pɒlɪs	door	dɜr
ice		nice	
paper	pebər	bag	begk
cream	krɪm	lipstick	lɪpstək
brush		cup	kɔrp
bottle	bɔrtəl	town	dɔn
government		England	
London		cinema	
picture hall		like	lek
number	nʌmbər	school	

The English language contains many words of Asian origin, including the following:

shampoo	bungalow	kedgeree	pyjamas
khaki	have a dekko	pukka	gymkhana

Do you think mother-tongue speakers of English would recognise these words as borrowings? Set up an experiment to find out.

RESEARCH
PATHWAYS

The language of bilingual speakers or writers in different situations or different generations; bilingual children in the early stages of language acquisition; original foreign language texts compared with their English translations; instructions, brochures, menus, or other written material which has been translated into English.

Website

There are many online resources for translation, foreign language learning, and etymology. Such sites as the dictionary pages referred to earlier (page 16) provide many links to machine translators, foreign language phrase books and etymological dictionaries.

The extent to which machine translation is possible is, in itself, a subject for research. If you know a language other than English or if you are studying one, you could compare the online translations of some selected phrases with your own versions. In particular, it is thought that machine translation cannot cope with non-literal language such as metaphor and idioms.

Bilingualism also features on pages concerned with language acquisition (see *Section B, Unit 6, Language acquisition*).

5 The language of social groups

The area of language research considers how language use may vary according to the social group a speaker (or writer) belongs to.

Human beings are very social animals; we live in groups of various kinds. Some different social groups include the following:

- Age
- Social class
- Gender
- Occupation
- Ethnic group
- Region

ACTIVITY

> Think about the social groups above, and write some notes on how you think membership of them might affect the language that an individual uses.

ACTIVITY

> If you have been working in groups, put the social group headings on a large piece of paper, and underneath each heading write the suggestions of all the small groups. These are your hypotheses (your predictions, your intuitions) about how language use might vary according to social group membership.
>
> Whatever your chosen method, the activities that follow will give you an opportunity to test out some of your hypotheses.
>
> First, read the notes on page 172.
>
> Then look again at your ideas on your previous brainstorm sheet. What aspects of language did you feel might vary according to the gender of the people involved?

Language and gender

The term 'sex' refers to the biological differences between male and female human beings; 'gender', on the other hand, refers to the extensive social conditioning that makes us think men and woman should be or behave in a certain way: for example, the fact that women can give birth to children distinguishes them biologically from men, and is therefore a sex difference; but the assumption that women are destined to look after children and clean the house, while men are not, is a gender construction. Much of our picture of how men and women should be is created by the language we have to describe and talk about the sexes.

ACTIVITY

To exemplify some of our possible expectations about male and female characteristics and behaviours, say whether the following pre-modifiers would be more likely to be applied to a male or a female figure:

athletic	bubbly	neurotic
well-built	emotional	rugged
elegant	kind	pretty
controlled	hysterical	nurturing
aggressive	sympathetic	independent

(from G. Morgan, 1986)

When you have finished this exercise, think about the following questions:

- Are there terms that are used of both sexes, but have a different meaning, according to which sex is being referred to? For example, does the term 'aggressive' have a more negative connotation when applied to women? does the term 'elegant' have a critical edge to it when applied to men? does the term 'well-built' mean the same thing when describing male and female bodies?

- For the terms that apply only to one sex or the other, is it the case that the other sex doesn't go in for the behaviour described, or is it that they do and the behaviour is called something else (including negative labelling, such as an athletic woman being called 'butch' and a nurturing man 'effeminate')?

In the exercise above, you have been studying the operation of connotation – that is, the associations that are built up in our minds between masculinity and certain traits, and femininity and other traits. These associations are thought to be the result of many different influences, including exposure to collocation patterns in language – i.e. the way in which some words habitually occur in the same environment as certain others. Sometimes collocations can be very fixed, such as in the phrase 'tall, dark and handsome', where not only do the words go together, but they go together in a particular order. But collocation can also refer to more loosely organised patterns, where linguists would talk about words having the *tendency* to occur with others. For example, Jennifer Herriman used a large corpus of language (the Cobuild Corpus of 50 million words) to search the modifiers that occurred to the left of 'man' and 'woman' in sentences. She found that almost half of the words were shared, denoting such factors as age and social class. But collocation patterns differed in, for example, men more often being described by occupation and women by appearance; also, women were more often described in terms of their marital status than men.

Connotation and collocation are active, evolving aspects of language use, and you can see the operation of these processes all around you. Here is an example for you to think about:

ACTIVITY

The cosmetics industry has witnessed huge growth recently as a result of the success of products aimed at men. According to *The Guardian* (3 December, 1999), the men's grooming market is now worth over £700m. The first UK men-only 'grooming and spa experience' has opened in London, called 'The Refinery'.

Here are the names of some of the different spa rooms inside The Refinery:

Turbo Boost

Pit Stop

Health Dock

The Body Shop has a male cosmetics range, including products called 'No Debate' and 'Activist'.

Explore the connotations of these new forms of product branding. How do they work to suggest particular ideas about masculinity?

ACTIVITY

If you focus only on branded advertising, you might be in danger of thinking that gender is all about what other people do, particularly those people in the media. However, we are all busily 'doing' gender in our daily behaviour, including everything we say and write, as part of the process of displaying ourselves. Display and advertising are different words for the same thing – putting on a show.

Read through the personal ads below, where male and female writers are advertising for partners. These ads are from a local free newspaper in the Derbyshire area.

Do you think the men and women involved are using language differently? If so, what are the differences in how they present themselves, and in how they represent possible partners?

What differences would you expect to see between the ads in a paper such as this and the ads in a national broadsheet paper, such as *The Guardian*?

The ads here represent a very small data sample. How big a sample do you think you would need in order to demonstrate patterns of usage in a convincing way?

MEN

Countryside fellow, own large bungalow, car, animals, pond, 45, 5ft 9, blonde, blue eyes, VGSOH. Seeks partner 30–50, who likes a challenge or something different – he (she) who dares wins!

Gay, good-looking non-scene male, 23, seeks similar age guy for friendship possible relationship.

Male bodyguard, 6ft, muscular build light brown hair and grey/green eyes, FT employment, OH. Looking for love, female any age/race, must be loving and caring and sincere as I am.

Stylish good-looking fellow, intelligent and caring, with lots of TLC to give. Seeks slim, attractive female for genuine relationship.

WOMEN

Cuddly young lady looking for gentlemen, 25–50 for friendship possible relationship.

Gay 20 yr old female smoker, looking for similar who is outgoing and aged 18–28 for nights out, friendship and plenty of laughs.

Lady, 38, 1 child, WLTM a businessman or farmer for friendship possible relationship, for having just general togetherness. Genuine replies only. Age and area unimportant.

Single lady, tall, slim with long auburn hair, green eyes. Likes the country, walking, gardens and places of interest, as well as travelling at home or abroad. WLTM sincere gent who has similar interests for friendship.

Note: OH = 'own house'.

ACTIVITY

Now read the extract below, which is from a 'Mills and Boon' novel entitled *Seduction* by Charlotte Lamb. The story is set in Greece, where the main female character, Clea, was to have an arranged marriage with a local Greek boy. However, the romantic hero, Ben Winter, arrives on the scene and attempts to seduce her. Clea is inexperienced sexually, and is torn between her need to please her father and accept the arranged marriage, and her desire for Ben.

In this chapter, Ben 'kidnaps' Clea from the beach after her early morning swim, and takes her to Athens for breakfast.

When you have read the extract thoroughly, make some notes on how the characters are depicted by the language that is used about them. *Note:* dots indicate points where the text has been edited.

Clea walked on towards the gate which let her out onto a stony, dusty lane. Beyond that the beach began ...

She swam for ten minutes, enjoying the salty spray which the wind flung into her face ... When she came out of the sea she halted in surprise ...

Ben was on his feet before she had turned away, his hand grabbed her arm, his fingers curled round her damp flesh in a grip which had no intention of being easily loosed. Startled, Clea lifted her head, her darting eyes wide, and met his little smile ... She looked down at her arm. The enclosing hand, darker in skin tone than her own, enforced an effortless grip on her ...

'Why won't you come?' Ben's voice had a sharp ring, the probe of his eyes fierce ...

'Will you let go of my arm, please? You're hurting!'

His fingers tightened rather than slackened, the grey eyes turning darker, filling with impatience. 'No, I'm not hurting you!'

He hadn't been, it was true, but now he was, and she sensed that he was doing it deliberately, his fingers biting into her. She looked down, trying to control a strange trembling which had begun inside her, in the pit of her stomach, as though she had swallowed a butterfly which was trying to escape ...

He opened the door of his white car as if to get into it, and Clea began to turn away. Hands fastened around her waist and she gave a muffled cry as she was swung up and round, deposited like a doll inside the car. Before she could get out again Ben was beside her in the driver's seat, the engine starting with a roar. Clea fumbled angrily at the handle as

continued

the car soared into flight, but Ben's arm shot out sideways and slapped her hands down from the handle.

'Let me out!'

'Sit still, and don't be a little idiot!'

She drew herself into a tight little corner, her eyes smouldering.

'You had no right to do this!'

'What have rights got to do with it? You wanted to come.'

'I did not!'

'Oh, yes, you did,' he mocked, his dark lashes covering his eyes yet leaving her with the distinct impression that he was watching her through them. 'You wanted to come as much as I wanted to take you.'

'If I'd wanted to come, I'd have accepted,' Clea denied.

ACTIVITY

If you have been working in groups, share your ideas both on how the male and female writers of the personal ads used language to describe themselves and others, and on how the male and female figures were depicted in the 'Mills and Boon' text.

Whatever your chosen working method, put together a written summary for your own purposes on the following further questions:

- Are there any similarities between the gender depictions in the different types of text? Give as many linguistic examples as you can.

- Do you think the way men and women are described in fiction affects the way we see ourselves in real life? Give some examples to support your opinions.

RESEARCH PATHWAYS

Look again at your hypotheses about how the factor of gender might influence language use, and at the notes entitled *Language and gender* (page 172). Choose one area, and try to gather some data to answer a question within that area.

Websites

There are as many different types of webpage that you could explore for gender as there are traditional, paper-based texts. For example, you could choose virtually any of the websites listed so far, from emoticons to newspapers, and investigate how gender is 'done' there.

In addition, you might research some online versions of the types of text you have studied in this section, and see whether there are differences. Here are some addresses for personal ads:

http://www.abcdating.co.uk
http://www.love-makers.com
http://www.geocities.com/Paris/Maison/2840
(specifically inter-racial ads)

Occupational register

Different occupations all have their own particular language which employees entering that profession have to learn.

Obviously, occupational registers are not entirely separate languages, although some professions – such as the legal profession – may use some forms of written language which are very difficult for the uninitiated to understand. An example of such a text in the legal profession would be a Will, or a Property Conveyancing Deed. In general, occupational registers may involve employees using commonly known words in a new sense – such as 'menu' in the computer industry, or 'poor' meaning 'not good' in the teaching profession; there may be uses of acronyms or other types of abbreviation, which assume shared knowledge by the users; whole structures of language, as well as single words, may be commonly preferred by people in the same profession; and there may be elaborated and detailed vocabulary associated with certain activities, artefacts or areas of experience.

ACTIVITY

Make a list of some different occupations, and give one or two examples for each of words and phrases you think are used in that occupation.

If you can't think of any examples from real life, then think about the many TV programmes that feature occupational life, and try to recall some examples from them. (Of course, you also need to assess how far these programmes are 'true-to-life', so if there are people in the group who have some knowledge of the real occupations, that would be useful for comparison.)

Here are some examples of programmes, to start you off:

| ER | The Bill | Vets in Practice | Peak Practice | Cops |

ACTIVITY

> If you have been working in groups, make a whole group list of
> occupations on a large sheet of paper, and, for each occupation, write
> up the terms you found.
>
> Whatever your chosen working method, consider the following question:
>
> - Why, in your opinion, do different professions use occupational
>   registers: what is their purpose?

ACTIVITY

> Read through the example of occupational register on page 179.
>
> When you have read it thoroughly, make notes on the following:
>
> - What *graphological* features are characteristic of teachers' report
>   writing, both from the evidence provided here and from your own
>   knowledge?
>
> - What conclusions could you draw from the *semantic* aspect of this
>   occupational register? Are certain words and phrases used here
>   characteristic of those often used in teachers' reports? Are certain
>   terms used with a specific meaning? Are there words and phrases
>   used that would be found less often in everyday uses of English?
>   What does studying the semantic level of language tell you about
>   teachers' professional concerns and preoccupations?
>
> - Are there particular *grammatical* structures used in the reports? What
>   are their effects?

ACTIVITY

> If you have been working in groups, pool your ideas on the linguistic
> features of teachers' occupational register.
>
> Whatever your chosen working method, answer the following additional
> *discourse* questions:
>
> - Who is the *audience* for the reports?
>
> - What is the *purpose* of the reports?
>
> - Occupational registers are subject to *change*. Is there any evidence,
>   from your knowledge of how teachers write their reports now, that
>   teachers' occupational register has changed? If so, why has it
>   changed, in your opinion?

DALEFORD HIGH SCHOOL

SUBJECT	GRADE	COMMENT	STAFF INITIALS
ENGLISH	B	Janet has improved tremendously throughout the year, keeping up a very satisfactory standard of work and effort, and producing some impressive assignments. Keep it up!	CN
MATHEMATICS	D	Janet's lack of application this year has brought its own reward in her poor examination result. She must try to concentrate more in lessons if she is to make any progress. She is not doing herself justice.	EK
FRENCH	B	Janet has a natural flair for this subject, and has demonstrated pleasing progress this year, particularly in her oral skills. She could perhaps pay more attention to the presentation of her written work at times.	RS
SCIENCE	D	Janet lacks interest and motivation, working inconsistently and often handing in homework that has been hastily executed. She needs more determination to succeed in a subject which she does not find particularly easy.	PK.
GEOGRAPHY	E	Janet's performance in this subject leaves much to be desired. Much more sustained effort is required, and less willingness to be distracted from work. She behaves immaturely at times, and her examination result was disgraceful.	KW.
HISTORY	C	Janet is always co-operative and friendly. However, her level of achievement does not reflect her true ability in this subject. She has considerable potential, but needs to prepare her work more thoroughly.	WO.
TECHNOLOGY	C	Janet is too easily satisfied with less than her best. If she made more effort, she would be able to achieve a better standard.	Kmm.

RESEARCH PATHWAYS

Look back to your original list of occupations, and examples of terms from the activity on page 177. Choose one of the occupations, and research the language that is actually used in that profession.

Website

If you are interested in the acronyms and vocabulary used in occupations, go to the following dictionary site and you will find acronym dictionaries and dictionaries devoted to particular special interests:

http://www.facstaff.bucknell.edu/rbeard/diction.html

The following will be useful if you want to explore the language of computing and the Internet:

http://www.netlingo.com

You can find more about this in *Section B, Unit 7, Speech and writing.*

6 Language acquisition

Language acquisition is concerned with how children acquire language as they develop as language users.

Unit 3 in *Section A: What do you know?* will already have made clear some of the different aspects of language competence that children have to acquire. It will also be clear from that unit that the area of language acquisition is very wide-ranging, so, for research purposes, it is necessary to focus on a manageable and discrete aspect of acquisition.

One division that it is possible to make for research purposes is that between the *features* and *functions* of language.

Children have to learn about the *features* of the language system they are acquiring, i.e. the different language levels. They also have to learn how these different systems combine to form recognisable genres of *speech and writing*. As part of this process, they have to realise that the two channels of speech and writing are, in themselves, very different in the way they work.

As well as acquiring knowledge about all of the above, children also learn that language is used for different purposes, or *functions*: for example, to give information, to play, to persuade, to control behaviour, and so on.

Research on how children acquire language may take one or more of the aspects above, and focus on one child with the intention of showing how that child's development compares with what textbooks regard as the 'norm' for that child's age; it may compare children of different ages or abilities in order to account for different levels of achievement; it may compare a child whose first language was English with a child for whom English is a second language; it may compare a child showing typical development with another whose development is impaired for a particular reason. Groups of children may also be studied in order to examine the effects of social factors such as gender or social class; teaching material aimed at children of particular ages or abilities may also be examined in order to decide what the writers' notions of acquisition are.

ACTIVITY

> Read through the data below, which is the speech of a child aged 2 years 4 months. She is reciting 'Incey Wincey Spider' for her older sister. Make notes on how far the child is able to produce the adult sound system: which sounds has she learnt, and which sounds are still causing her some difficulty?
>
> Remember that this is connected speech, rather than words in isolation, so adults saying this rhyme may use features such as liaison and elision (see *Decoding spoken texts*, page 33).
>
> ɪsɪ wɪsɪ paɪdə kaɪmd ʌp də wɔtə paʊt daʊ keɪ de weɪ daps m wɔst pu
>
> wɪsɪ aʊt dʊ keɪ de sʌsaɪ daɪd ʌp ɔ də weɪ ɪsɪ wɪsɪ paɪdə kaɪmd ʌp
>
> dæt paʊt əgeɪ

ACTIVITY

> Read through the conversation below, which is between three 10 year-old children in a primary school classroom. The children have been asked by the teacher to design a board game. Originally, this game was to be based on their ongoing topic – 'Journey to the Centre of the Earth' – but the children decided they would like to design games for other classes in the school.
>
> The children devised 'Market Research' questionnaires, and having collated this information on the computer, they now have to make the final decision about the type of game they are going to create.
>
> No teacher was present during the conversation.
>
> When you have read the transcript, make some notes on the following:
>
> • What are the children using language for, on this occasion?
>
> • What features of language mark the dialogue as serving a particular purpose?
>
> A: Right, um ...
> B: Is it ready?
> C: Yeah, it's ready. Right ... so ... what have we got to do now?
> A: Right, we've gotta try and um ...
> C: The best one is skill and adventure ...
> B: They like adventure and skill the best. We need a skill and adventure game ...
> C: We could do a game like 'Hero Quest' where you design things like monsters ... like monsters ...
> A: Gotta be skilful ... yeah ... so we gotta design our game

continued

B: Shall we do it about our topic or shall we do it about their topic ... that they're doing about ... like skeletons and stuff?

C: Skeletons are more frightening ... or shall we do sort of like polar bears and stuff like that or ...

A: Polar bear, I mean, what?

C: Well – don't know ... some ... but sort of ... aliens

A: It's supposed to be adventurous like chance ...

C: Aliens are scary ...

A: Skeletons we could do about ...

B: That's not adventure and skill, is it?

A: Yes but they said that the scarier it is the more they like it ...

B: Yes, I know, but ... I mean – you're not gonna have skill and adventure ... the scary game and you say let's have skeletons stuck in the middle of the board ... I mean ... it's not very scary, is it?

C: You could search for the skeletons ...

B: No ... what you do is you have a game so you have a game which is really horr ... horrific ... that's like an adventure but you've got to go through and you've got to decide which one to go in like a big maze ...

C: A maze! That's it! A maze game!

A: No ... 'cos that's boring – not very easy, is it?

C: Have counters and you could make squares ...

A: It's not scary ...

C: You could say miss a turn ...

B: Not like ... not like a normal maze ... like a really ...

A: No, we're not doing a maze ... it's boring, that ... think of something else

B: It's boring ... how would you like it if you're just moving a counter up and down round a maze ...

A: It's gonna be exciting ... you could have your head chopped off by axemen

C: And you miss turns and go back to the start ...

A: Wow!

C: And you get experience that girl said ... remember ... like that girl said ... you get experience ...

B: Shall we do that, then?

A: Well ... I think we should have a vote

C: We don't need a vote ... we need to des ... why don't we design the two different ideas and the one that's the best gets done

B: Let's do ... all do an adventure ... well a maze

A: And decide which is the best ...

C: Not a maze, then ...

A: No ...

B: Alright ... an adventure game ... don't shoot, don't shoot!

A: An adventure game and we'll decide which one is best.

ACTIVITY

If you have been working in groups, share your ideas on the spoken data you have been studying.

Whatever your chosen working method, devise a way to present your findings in an accessible and economic form, so that patterns of language use are highlighted in analysis.

When you have done this, discuss what further data you could collect if you wished to undertake a *comparative* study of children's acquisition of spoken language.

ACTIVITY

Below are three examples of early writing from different activity areas in a nursery school. The writers of these texts were all approximately three and a half years old. *Sample A* is from the telephone area; *Sample B* from the hospital reception area; *Sample C* from the shop area.

Make notes on the following:

• What do these writers already know about writing?

• How far do the samples represent different genres of adult text?

SAMPLE A

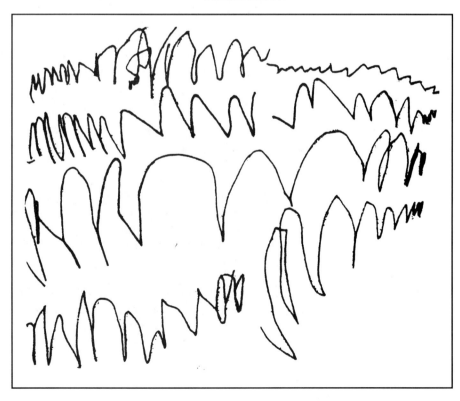

SAMPLE B   **SAMPLE C**

ACTIVITY

Read through the two stories on pages 186–187, which were both written by seven year-old children.

Make notes on the following:

- How much do these writers know about the structure of narrative texts?

- Analyse the writers' use of vocabulary and grammatical structures.

- Analyse the writers' knowledge of spelling patterns: where the writers have not quite matched adult spelling patterns, is there any rationale for the child's attempt, in each case?

SAMPLE 1: THE TOOTH FAIRY

One day I lost my tooth at scool
that nihgt I put it atr my pillow
and went to bed I'd forgotton my tea
Suddley a fant l'ght showed
it was marigold I got out of bed
and flew away with her and had tea
With her the hous was batiful it was made of
teeth then I went home

SAMPLE 2: PIRATE RED SHIP

Pirate red Ship

once there was a pirate
called pirate red ship he was
the leder and all the
crowe where called the
red ship and . they had
a red ship and
one day there was
a hyuwge crawd
arawend the red ship
the pirates were
going . to sea that
morning but at night
there was a storme
and there ship sunck
and they drawnt and
tide

ACTIVITY

If you have been working in a group situation, share your ideas on the narrative skills of the young writers, and of their knowledge of the writing system.

Whatever your chosen method of working, consider the following:

• How can you present your findings in a way that gives due credit for what the young writers have achieved, rather than what they have yet to learn?

• How could you conduct a *comparative* study of the development of children's writing skills?

RESEARCH PATHWAYS

Look again at the initial notes on language acquisition (page 181). Choose one area listed, and consider how you could collect data for that.

Websites

You will find material on language acquisition on many sites where general language issues and questions are being explored. For example, the 'bucknell' dictionary site on page 180 has a resource called 'the human languages page'; many homepages of commercial dictionary publishers also have links to resources of this kind. For an example of the latter, try:

http://www.m-w.com
(this is Merriam Webster's homepage).

If you want to explore bilingualism, try:

http://www.familyculture.com/bilingualism.htm

7 Speech and writing

This area is concerned with comparing spoken and written language.

It will be clear from the material in *Section A* on written and spoken texts that the systems of speech and writing are very different in nature. But the channels of speech and writing are also very different in function, in what they are used for.

ACTIVITY

Look at the headings below. Try to give six examples for each column.

What is speech good for?   *What is writing good for?*

ACTIVITY

If you have been working in groups, share your ideas on the functions of speech and writing by collating your results on a large sheet of a paper.

If you have been working individually, ask three informants to complete the same exercise for you, then collate the results.

To what extent are the functions of speech and writing different?

ACTIVITY

If you are going to compare spoken and written texts, you need to find situations where the *content* is likely to be the same.

You will have realised, from doing the previous activity, that we use speech and writing for very different purposes – so finding good comparisons is not as easy as you might think.

Look at the possible comparisons on page 190.

Which of these would provide good comparisons for spoken and written language use, and which would not?

continued

1. A football commentary with a match write-up in the Press.

2. A leaflet from a DIY store explaining how to lay carpet tiles, compared with an in-store video explaining the same.

3. A recipe from a book compared with a recipe from a cookery programme on TV.

4. A teacher's delivery of a lesson, compared with her written documentation for that lesson.

5. A novel, compared with the filmed version of the novel.

ACTIVITY

Go through your findings, taking clear notes for your own record, to decide exactly why some of the comparisons would not yield fruitful results – what would be the problem?

ACTIVITY

Read through the two pieces of data below and on page 191. *Text A* is the transcript from a weather forecast given by John Hammond after ITV's Nightly News; *Text B* is the weather forecast from *The Times* for the same day.

When you have read the material thoroughly, consider some of the ways in which the two versions differ.

TEXT A

Good evening/a frost-free spell of weather just at the moment/a very mild night tonight/in most places around about eight or nine/but ah up in the far north east of Scotland/two/and that's cold enough for some sleet or snow/across parts of Orkney and the far north east of the mainland/you'll see some rain edging into the north western parts of Scotland and Northern Ireland during the night/elsewhere mostly dry/the odd patch of drizzle/especially in the west/a lot of cloud out there/fairly breezy/a fairly breezy start tomorrow and a good deal of cloud again/we'll see some patchy rain travelling across Scotland and Northern Ireland/into England and Wales but turning very light by the time it reaches the south/some places staying dry/and it'll brighten up across eastern parts of Scotland/north eastern parts of England/helping temperatures up as high as twelve or thirteen degrees/fifty five fahrenheit/and that wind will ease off for a time/but then picking up again later on in the far north west/gales returning here/the rest of the week looks very mild/temperatures remaining in double figures/just about everywhere/some brightness/but a lot of cloud/with some patchy rain from time to time/especially in the north and the west/that's about it from me/I'll leave you with the summary

© Crown copyright

TEXT B

FORECAST

General Situation: England and Wales will have a breezy, fairly cloudy day with some rain. Most of the rain will be in the NW, with the south and sheltered east mainly dry with bright spells. Scotland and N Ireland will have thick cloud bringing outbreaks of rain. Over N Ireland and eastern and central Scotland there will also be long, dry periods.

London, SE, E, Central England, E Anglia, Midlands, Channel Islands: Cloudy with some bright spells. Mostly dry. Wind: SW moderate or fresh. Max 11C (52F).

SW, NW England, Wales, Lake District, Isle of Man, SW Scotland, N Ireland: Cloudy with patchy rain. Some dry and locally brighter spots. Wind: SW fresh. Max 10C (50F).

NE England, Borders, Edinburgh & Dundee, Aberdeen, Moray Firth: Mostly dry and bright. Wind: W or SW, strong and gusty. Max 11C (52F).

Glasgow, Central Highlands, NE, NW Scotland, Argyll, Orkney, Shetland: Dull with rain, some heavy. Wind: W or SW strong locally gale. Max 9C (48F).

Republic of Ireland: Cloudy. Rain at times. Wind: SW fresh and gusty. Max 11C (52F).

Outlook: Remaining mild, windy and unsettled. Most of the rain in the north.

AROUND BRITAIN

Yesterday: b=bright; c=cloud; d=drizzle; ds=dust storm; du=dull; f=fair; fg=fog; g=gales; h=hail; r=rain; sh=shower; sl=sleet; sn=snow; s=sun; t=thunder

	Sun hrs	Rain in	Max ¡C	¡F			Sun hrs	Rain in	Max ¡C	¡F	
Aberdeen	1.6	0.01	8	46	c	Jersey	9.3	-	10	50	s
Anglesey	6.6	0.01	10	50	s	Kinloss	1.7	0.01	9	48	r
Aspatria	5.6	-	9	48	s	Leeds	6.8	0.01	10	50	b
Aviemore	2.3	0.04	6	43	r	Lerwick	0.0	0.31	6	43	r
Belfast	2.7	0.01	9	48	du	Leuchars	0.4	0.01	7	45	r
Birmingham	4.5	-	9	48	b	Littlehampton	6.7	-	-	-	s
Bognor R	7.8	-	10	50	s	London	4.3	0.01	10	50	s
Bournemouth	7.7	-	10	50	s	Lowestoft	5.3	-	8	46	b
Bristol	8.1	0.01	10	50	s	Manchester	5.5	0.02	10	50	s
Buxton	4.9	0.04	7	45	s	Margate	0.7	-	9	48	s
Cardiff	7.2	0.02	10	50	s	Morecambe	5.8	0.01	9	48	s
Clacton	7.7	-	8	46	s	Newcastle	4.8	-	10	50	s
Colwyn Bay	6.4	-	10	50	b	Newquay	8.3	-	10	50	s
Cromer	5.8	-	8	46	s	Norwich	5.7	-	9	48	s
Edinburgh	-	-	8	46	r	Oxford	7.7	-	10	50	s
Eskdalemuir	1.4	0.03	7	45	r	Penzance	7.6	-	11	52	s
Exmouth	9.6	-	10	50	s	Poole	8.5	-	10	50	s
Falmouth	7.1	-	12	54	s	Prestatyn	7.1	-	9	48	s
Fishguard	5.9	0.01	9	48	s	Ross on Wye	7.2	-	10	50	s
Folkestone	4.4	-	9	48	b	Saunton Sands	8.2	-	12	54	s
Glasgow	-	-	7	45	r	Scarborough	3.1	0.01	8	46	b
Guernsey	9.8	-	11	52	s	Shrewsbury	6.6	0.01	10	50	s
Hastings	-	-	10	50	s	Skegness	5.4	0.01	9	48	s
Hayling I.	6.4	0.01	10	50	s	Southend	4.0	-	8	46	s
Herne Bay	5.6	-	9	48	s	Southport	6.7	-	9	48	s
Hunstanton	5.6	-	9	48	b	Stornoway	0.0	0.15	10	50	r
Isle of Man	5.3	-	9	48	s	Swanage	9.0	-	10	50	s
Isle of Wight	7.2	0.01	10	50	s	Teignmouth	9.5	-	10	50	s

Yesterday: Highest temp: Falmouth (Cornwall) 12C (54F); lowest max: Aviemore (Highland) 6C (43F); highest rainfall: Lusa (Isle of Skye) 1.43ins; most sunshine: Guernsey (Channel Islands) 9.8hrs.

ABROAD

Ajaccio	15	59	s	Cologne	4	39	sh	Madrid	16	61	s	Rome	17	63	s
Akrotiri	17	63	s	C'phagn	3	37	s	Majorca	17	63	s	Salzburg	0	32	sn
Alex'dria	18	64	s	Corfu	15	59	f	Malaga	18	64	s	S Frisco	13	55	f
Algiers	17	63	s	Dublin	8	46	f	Malta	16	61	s	Santiago	28	82	s
Amst'dm	5	41	f	Dubrovnik	11	52	s	Melbourne	21	69	f	S Paulo	31	88	s
Athens	18	64	s	Faro	15	59	s	Mexico C	10	50	f	Seoul	10	50	f
Bahrain	20	68	s	Florence	13	55	s	Miami	26	79	f	Singapore	32	90	f
Bangkok	34	93	f	Frankfurt	5	41	f	Milan	14	57	s	Stockholm	-4	25	f
Barbados	29	84	sh	Funchal	16	61	c	Montreal	1	34	s	Strasb'rg	5	41	f
Barcelona	15	59	f	Geneva	4	39	s	Moscow	2	36	f	Sydney	24	75	f
Beijing	13	55	s	Gibraltar	18	64	s	Munich	1	34	sn	Tangier	24	75	s
Beirut	17	63	s	Helsinki	-5	23	sn	Nairobi	26	79	f	Tel Aviv	18	64	f
Belgrade	5	41	f	Hong K	26	79	s	Naples	17	63	f	Tenerife	24	75	s
Berlin	3	37	f	Innsbruck	4	39	s	N Delhi	25	77	fg	Tokyo	12	54	f
Bermuda	19	66	s	Istanbul	13	55	f	N York	8	46	s	Toronto	7	45	c
Biarritz	10	50	s	Jeddah	28	82	s	Nice	15	59	s	Tunis	18	64	s
Bordeaux	9	48	s	Jo'burg	21	70	f	Oslo	1	34	s	Valencia	17	63	s
Brussels	6	43	f	L Palmas	24	75	c	Paris	7	45	s	Vanc'ver	8	46	f
Budapest	4	39	s	Le Tquet	9	48	f	Perth	31	88	s	Venice	10	50	s
B Aires	30	86	f	Lisbon	12	54	s	Prague	1	34	sn	Vienna	3	37	sn
Cairo	20	68	s	Locarno	12	54	s	Reykjavik	4	39	sh	Warsaw	1	34	sn
Cape Tn	28	82	s	L Angeles	13	55	c	Rhodes	17	63	f	Wash'ton	12	54	s
Chicago	13	55	f	Luxembg	4	39	s	Rio de J	34	93	s	Wel'ngton	17	62	f
Ch'church	20	68	s	Luxor	25	77	s	Riyadh	23	73	s	Zurich	3	37	s

Temperatures at midday local time yesterday. X=not available

NOON TODAY

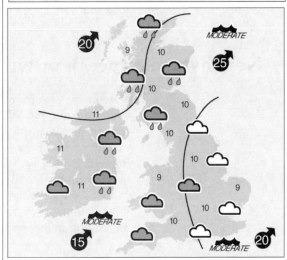

Changes to the chart below from noon: Low D will edge eastwards and fill. Low P will move northeastwards and fill with high pressure remaining to the south of the UK.

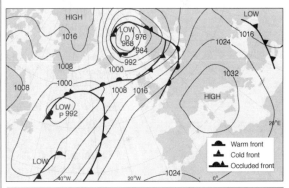

HIGH TIDES

TODAY	AM	HT	PM	HT	TODAY	AM	HT	PM	HT
Aberdeen	01.35	4.1	13.38	4.2	Liverpool	11.36	9.4	23.55	9.3
Avonmouth	07.25	13.1	19.44	13.1	London Bridge	01.57	7.0	14.21	7.2
Belfast	11.19	3.4	23.36	3.3	Lowestoft	09.56	2.4	21.50	2.5
Cardiff	07.11	12.1	19.30	12.1	Margate	-:--	--	12.27	4.5
Devonport	06.00	5.4	18.21	5.3	Milford Haven	06.27	6.8	18.44	6.7
Dover	11.14	6.4	23.34	6.6	Newquay	05.18	6.8	17.35	6.8
Dublin	11.41	4.0	-:--	--	Oban	06.05	4.0	18.20	3.8
Falmouth	05.23	5.2	17.47	5.0	Penzance	04.48	5.4	17.05	5.3
Greenock	00.34	3.0	12.56	3.0	Portland	07.05	2.1	19.35	2.2
Harwich	-:--	--	12.10	3.8	Portsmouth	11.30	4.5	23.56	4.5
Holyhead	10.33	5.6	22.54	5.4	Shoreham	11.30	6.0	23.51	6.2
Hull (Albert D)	06.35	8.3	18.41	8.7	Southampton	11.08	4.4	23.35	4.5
Ilfracombe	06.14	9.0	18.30	9.0	Swansea	06.35	9.2	18.50	9.2
King's Lynn	06.41	6.4	18.41	7.0	Tees	03.57	5.1	15.58	5.3
Leith	02.52	5.3	15.02	5.4	Walton-On-Naze	-:--	--	12.06	4.1

All times GMT. Heights in metres.

ACTIVITY

If you have been working in groups, share your ideas on how and why the spoken and written forecasts differ.

Whatever your chosen working method, write up your analysis fully, using the headings that are appropriate for the data, so that any interested reader would find your ideas accessible.

RESEARCH PATHWAYS

> Go back to the list of comparisons in the activity at the bottom of page 189, and try to add further comparisons to this list. Choose one comparison, collect data for this, and analyse it.

New hybrids: speech or writing?

New forms of communication, such as computer-mediated communication (CMC), are blurring the distinctions between speech and writing. For example, in Internet chat-rooms, people write to each other in real time, so they experience the fast give-and-take and the immediacy that we would normally associate with speech. But there is no voice in the traditional (i.e. acoustic) sense; and the participants cannot see each other, so there is no way of getting the kind of social information – such as age, sex, ethnicity – that we would normally be registering in face-to-face encounters.

ACTIVITY

Below is a chat-room encounter between participants who know each other in real life (Xerxes is Orc's neighbour and they share duties such as cat-feeding when either of them is away from home).

The names Xerxes and Orc are log-in names; these are names that users adopt so that they do not have to reveal their real identities online.

In what ways is this dialogue like a spoken interaction, and in what ways is it not?

You might also ask the converse: in what ways does this text follow the rules of writing, and in what ways does it not?

```
Xerxes: hi, didn't expect you here!
Orc: you never know who you are going to meet in cyberspace
Xerxes: how did you know I'd be here?
Orc: we always look out for U my dear!
Xerxes: well I'm hanging around to see if my friend Marie
appears (from Sweden)
Xerxes: cat's fine, btw, she's being cool...
Orc: is she wearing shades?
Xerxes: yeah, hipcat wraparounds
Orc: nothing but the best for ella
Orc: is your Swedish friend on aol then?
Xerxes: yeah
Orc: that's handy then
Xerxes: I've just got the latest version of netmeeting,
should be neat
Orc: I think S wants to talk to you about web pages and
setting uup chat rooms at some point
Xerxes: fancy getting a camera? ok, fire away!
Xerxes: looks like Marie isn't coming
Orc: *gone for more beer* - i htink it would be best to chat
```

continued

```
in person
Xerxes: ok...what's happening in Glasgae,
then?
Orc: i have had a relaxing day - reading etc - then we went
out for a meal
Xerxes: what are you doing online anyway?
Orc: i was just checking my mail
Orc: i got an email from C which didn't go down too well
Xerxes: so did you come into this space or were you somewhere
else?
Xerxes: what did it say?
Orc: just asking what i was up to for my birthday
Xerxes: which is when?
Orc: when you sent me the message (march 29) a window asked
if i would except the message from you
Orc: you can set up a buddy listr so when people you know log
on it lets you know
Xerxes: ok I get it..so what are you doing for your b-day?
Xerxes: I think that's what I did, that's why you showed up
on my list
Orc: don't know yet -its a wednesday so a couple of drinks in
town maybe - you are welcome to join in if we are going to
do somehting
Xerxes: ok! I'm going to go off now and send Marie a mail to
see if she's around
Orc: sorry - we are going to have to continue our chat
another time -
Orc: bedtime calls
Xerxes: CYAL8R!
Orc: bye
Xerxes: tarra luv!
```

RESEARCH PATHWAYS

Studies of the differences between chat-rooms and face-to-face conversation; studies of computer bulletin boards such as those of usenet groups; analysis of the nature of e-mail communication; the nature of online 'published' texts (such as public information sites, advertisements, university homepages, etc.); the metaphorical nature of language in the computer environment (for example, 'desktop', 'mouse', 'trash can', etc.); magazines advertising computers and associated products.

Website

If you want to research the specialist language surrounding computers, try:

http://www.netlingo.com

You should be aware that researching aspects of CMC poses particular ethical problems. For example, it is unethical to take someone's posting or chat-room conversation and use it without permission. Additionally, researchers in this area are often members of the group they are researching (a technique called 'ethnography'). So while you can study some aspects of computer language at a remove (e.g. the metaphors above) you cannot study people's computer-based communications in the same way.

8 Interaction analysis

Interaction analysis (or discourse analysis) is concerned with how language is used in encounters between people.

The focus for a study of interaction may be a general one, for example, how the participants manage the conversation between them; or it may be more specific, for example, how a certain individual controls the dialogue by the linguistic strategies he or she uses.

Depending on the focus for the study, data collected may consist of one dialogue, or more than one. If a researcher wanted to compare the rules in different types of interactions, for example, the rules of a formal debate, compared with an informal conversation, more than one type of spoken data would need to be gathered.

ACTIVITY

Look at the examples of possible interactions:

- A televised debate

- Language use in the courtroom

- Children or teenagers talking in peer groups

- Language use in the House of Commons

- A teaching situation

- A radio phone-in programme

- A telephone sales conversation

- Door-to-door sales conversations

- A church ceremony

- A family conversation

What questions might be of interest to a researcher studying the language use in these situations?

ACTIVITY

ACTIVITY

Read through the dialogue below, which is a transcript of a Brownie meeting.

Discuss how Muriel, the Brown Owl, controls the interaction between herself and the Brownies: what discourse strategies does she employ?

M = Muriel   K = Katie   J = Joanna   E = Elizabeth   C = Caroline
S = Samantha   L = Louise   A = Ann   R = Rachel   N = Nicola
T = Tawny Owl

TRANSCRIPT FROM A BROWNIE MEETING

M: Right Brownies, well just looking round a few Brownies tonight would have lost points and does anybody know why they would have lost points? For inspection we are talking about.

M: Katie?

K: Shoes?

M: Yes. Some of them haven't brought hats but some haven't got trainers on or pumps and you'll automatically be penalised and you will lose points. And that Samantha and Jane and anybody else and Joanna is because we can't have you running in shoes or sandals where you'll slip. You must have trainers or pumps ... so will you, just wait a minute Caroline ... so will you remember for next week ... As it happens we can't find the books and we're not having inspection tonight so ... but for next week. Have you got pumps? Good. Have you Samantha? And how about Joanna, you've got trainers?

J: Yea.

M: Oh well that's alright then ... fine, right would you just like to sit down and we'll have a pow-wow.
     (Noise)

M: Right now Brownies would you like to tell me what you've been up to and where you've been on your holidays because a lot has been happening and it's a long, long time since we all had a pow-wow. First of all Elizabeth.

E: I went to Alton Towers ... you know on the corkscrew.

M: Oh and what did you do when you were there?

E: Well um we didn't go on the corkscrew ... we went to, we were

continued

meaning to go to the circus but when it was time to go we were somewhere else dead far away from ... so we didn't go and er we went on the ferry ... and oh yes ...

M: And what else?

E: The Magic Carpet and Michael started crying because he couldn't go on because you have to be a certain height and I could go on but he would have been crying when he came off it was awful.

M: Was it scary?

E: Yes.

M: Was it very high?

E: Well it was like that ... like that.
     (Cough)

M: Right Joanna, hands down the rest of you we'll give you all a turn.

J: We went to see my Auntie E ... in the ... and er she used to live next door to us but she's moved to live with her husband er in Wales and er when we got there we went to Rhyl and when we came back she started crying. Because er we went to see her because her husband's in hospital he's only got three months to live.

M: Er how old is he?

J: Er I don't know.

M: Not very old is he ... oh dear. Oh that is sad ... but did you have a nice holiday? Good, Caroline.

C: The best day of my holiday was my birthday.

M: Oh it would be. Tell us what you did on your birthday.

C: Nothing much because it was pouring down.

M: Well why was it the best day then? What presents did you get?

C: I got Peaches and Cream Barbie.

M: Who?

C: Peaches and Cream Barbie.

M: Is that a doll?

C: Yea I got a skirt, two tops, two pairs of knickers.
     (Laughter)

M: Right now Samantha.

S: ...

M: Can you speak up love because I've got very bad hearing today. Where did you go to?

S: Chester.

M: Chester ... Chester Zoo? Oh just Chester and what did you think about Chester?

S: ...

M: We like it, don't we Tawny? What did you like best?

S: The River.

M: The River, did you go on the river?

S: And ...

M: And did David go?

S: ...

M: Did he like it? Yes. Did your Daddy like it? Good how about Louise?

L: I went to see the Minster at York.

continued

M: Oh.

L: ...

M: Well tell the Brownies what the Minster is because some of them might not know what the word means.

L: Yea.

M: It's another word for a large church.

L: There was all ...

M: It's like a cathedral. Good. Now, who's next? Ann.

A: I went on holiday and went to York.

M: Oh York ...

A: We went inside York Minster then went in the Tower ... down the road from the camp site was a little town full of shops ...

M: Was it nice?

A: Yea ... a penny arcade and they had all old machines in it.

M: Oh lovely.

A: And the lady let us ... special coins to have a go.

M: That's right, special tokens. Now then, we'll move round this way next. Laura? Oh dear me, Rachel? I'm thinking of ... right Rachel go on.

R: I went to the Lake District ...

M: The Lake District? Was it good and did you get some good weather? No, did it rain? Oh dear. But you enjoyed it? How about Michelle? No, Elizabeth.

E: When we was walking back from the pictures we saw this dead cat lying on the grass.

M: Oh my goodness. Right I don't want any more tales of dead cats. Can we have holiday news please. Right, how about our newcomer, what's your name love?

N: Nicola.

M: Nicola, right Nicola you tell us what's ... where you've been love.

N: Wales.

M: Wales, where about in Wales.

N: Somewhere in the middle.

M: ... in the middle.

N: Yea.

M: Oh Mr Jones, did you have a nice time?

N: Yea, it was my birthday when we went.

M: Did you go for a week?

N: No a month.

M: A month.

N: My Grandad and all my Aunties and all my cousins live up Wales.

M: Oh aren't you lucky.

N: Yea.

T: ... Evans.

M: Oh Evans oh well that explains it. Right, what it is a lot of people live in Wales they are all called Evans or Jones that's why it's funny. Anyway Samantha.

S: ...

M: Can you speak up love I can't hear a word.

ACTIVITY

If you have been working in groups, share your ideas on the strategies used by Muriel in talking to the Brownies.

Whatever your chosen working method, do the following:

- Decide why this conversation is humorously typical of those that arise in teaching situations – which particular discourse strategies are carried to extremes in this dialogue?

- Make a list of discourse strategies as headings for your analysis, then write a summary of your findings under each heading. Organise your work so that any interested reader would find it accessible.

RESEARCH PATHWAYS

Add to the list of 'speech events' you studied in the activity on page 194. Then choose one of the events, and decide what question you want to ask about it. Collect some data that will enable you to investigate that area with the initial question in mind. If the data does not answer the original question you posed in enough detail, are there other questions that are highlighted by the data you have collected? Make some appropriate headings for your analysis, and write it up in a readable and lively way.

Section C

TAKING STOCK

1 Planning your investigation

This final section will be particularly useful to you if you are thinking about starting a language investigation. It doesn't matter how far into the future you might be looking: although some people don't think about the stages of research or its assessment until their research is underway, these are invariably the people who don't do very well. It's important to know what you are aiming at from the outset.

Research involves a process as well as a product.

If you understand and carry out the process of research as outlined below, the product you create will be a satisfying outcome to all your efforts. The best pieces of research always arise from thoughtful consideration and negotiation between researchers and supervisors.

Poor research invariably tries to short-cut the process, avoiding consultation and not using supervisors as a resource.

The stages of production

Stage 1
Discuss with supervisor the area you would like to research.

Agree that you will collect some data by a specified date.

Stage 2
By negotiation with supervisor, decide on a clear question you are going to ask about your chosen research area, after careful consideration of the data you have collected. Select data, if you have collected more than you need for the question you are asking.

Stage 3
Read any appropriate secondary sources. Make notes. Remember that secondary sources are a guide to the area you are researching, rather than being the final word on what you should find: you may well find all kinds of details not mentioned by academic research. Don't be blinded by what textbooks have to say, and don't worry if there's not much written on the area you are researching.

Stage 4
Create a plan with your supervisor. A plan involves setting up a number of headings for analysis of your data, so that you will cover all the areas necessary in as comprehensive a way as possible.

Stage 5
Consider permissions and confidentiality: if your material is spoken data, get permission from the speakers to use it; if your data is written material of a personal nature, ask the owner/writer for permission for use.

Stage 6
Write out transcripts, if the investigation involves spoken data; decide how you are going to present your data for the reader.

Stage 7
Write your introduction, explaining to the reader what you intend to do.

Outline the method(s) you have used to collect data.

Stage 8
Start your analysis, working through each heading in turn.

After completion of each section, check with your supervisor that you are on the right lines. It is not a supervisor's job to return your work 'corrected', but rather to give you feedback on whether you are analysing your data in enough detail. He or she may give you one or two examples of what this means in practice, then the rest is up to you. As you are working on your analysis, check that you are presenting your ideas in the most economic way: for example, could your findings be presented in tabulated form, rather than continuous prose? Keep your word count in mind as you work.

Stage 9
Write your conclusion. This should not be a repetition of large chunks of your investigation, but a summary of your main findings. Think about what you want your readers to remember, after they have finished reading your work – what is the main message you want them to take away?

Stage 10
Complete the final write-up. Check presentation, and secretarial aspects of your writing.

ACTIVITY

Read through the stages outlined above, and look at a calendar or diary to see how much time you will have for your research. Prepare a list, under the headings *Stage* and *Date*, and put a deadline date against each of the ten stages to help you to plan your work. Keep this at the front of your research file.

2 Assessment criteria

This unit provides a checklist of the general criteria that are used by supervisors when they mark investigations.

Although research at different school key stages, for A level specifications and on Higher Education degrees will vary in its level of sophistication, supervisors invariably look for certain common qualities in the work produced. Of course, you will need to supplement what is written here with any more specific assessment guidelines given to you for the particular course you are taking.

If you have past copies of investigations available, it would be a useful exercise to 'mark' some of these yourself, in order to get a practical sense of what these criteria mean in practice. Whether past investigations are available or not, it will be useful to ask yourself these questions at various points in the research process.

1. Ability to ask meaningful questions about language
Is the question you have asked a useful one?
Does it make sense?
Is it a reasonable question to ask?

2. Ability to arrive at and maintain a focus, purpose, and sense of direction
Does the investigation know where it is going?
Does it progress in a logical fashion, without being vague or going off at irrelevant tangents?
Does it stay on target, and keep the original question in sharp focus?

3. Reasonable scope
Does the investigation set itself a reasonable task, or is it trying to do far too much at once?

4. Good data
Is your data good for the question you are asking, or does it not have enough in it to enable you to say very much?

5. Open-mindedness
Have you been honest in the way you have collected your data?
Have you really looked at your data, or did you have preconceived ideas about what you should find, and therefore tried to make your data fit your foregone conclusions?

6. Rigorous, accurate and thoughtful analysis
Have you analysed your data in detail, or only commented superficially on it?
Have you analysed (i.e. said *why* the language was used as it was) or only described what was in the data?
Have you looked for patterns of usage in your data?
Is your analysis accurate?
Did you give some thought to how best to analyse your data?

7. Degree of engagement with material
Have you really quarried your material in your analysis, or have many things been left unsaid?
Have you shown enthusiasm for your chosen task and interest in the data you collected?
Were you motivated and determined to find answers to the questions you originally posed?

8. Evidence of learning
Is there any evidence that you discovered some things you didn't know before?

9. Perseverance and initiative
If you met difficulties, did you persevere, and try to think round the problems, or just give up?
Were you resourceful in how you went about your work?

10. Readability and word count
Is your style readable, clear and economic?
Have you stayed within the word count?
Has your investigation been presented in a way that makes it accessible to the reader?

11. Relevant and realistic conclusions
Are your conclusions clearly related to the investigation itself?
Have you claimed to have done things that you haven't?

12. High level of technical accuracy
Have you made many errors in expression, spelling and punctuation?

THE LEARNING CENTRE
TOWER HAMLETS COLLEGE
ARBOUR SQUARE
LONDON E1 0PS

References and further reading

Crystal, D. and Davy, D., (1980). *Investigating English Style*. Longman.

Eggins, S. and Slade, D. (1997). *Analysing Casual Conversation*. London: Cassell.

Grice, H. P. (1975). *Logic and Conversation*. Reprinted in P. Cole and J. Morgan (eds.) *Syntax and Semantics, Vol 3: Speech Acts*. New York: Academic Press.

Herriman, J. (1998). 'Descriptions of "woman" and "man" in present-day English'. *Moderna Språk,*. XCII(2), 136–42.

Milroy, J. and Milroy, L. (1993). *Real English: The Grammar of English Dialects in the British Isles*. Essex: Longman.

Morgan, G. (1986). *Images of Organization*. London: Sage.

Further reading

There are several accessible series of textbooks that cover a range of different topic areas and skills. Details below:

The Intertext series (Routledge) contains a core book on language analysis, entitled *Working with Texts*, plus a number of smaller topic books, including the following:

Beard, A. (1998). *The Language of Sport*.

Beard, A. (1999). *The Language of Politics*.

Goddard, A. (1998). *The Language of Advertising*.

Goddard, A. & Mean-Patterson, L. (2000). *Language and Gender*.

McLoughlin, L. (2000). *The Language of Magazines*.

McRae, J. (1998). *The Language of Poetry*.

Reah, D. (1998). *The Language of Newspapers*.

Ross, A. (1998). *The Language of Humour*.

Sanger, K. (1998). *The Language of Fiction*.

Shortis, T. (forthcoming). *The Language of ICT: Information and Communication Technology*.

Other useful series include *Living Language* (Hodder and Stoughton) and *The Language Workbooks* (Routledge).

Index

This index does not list every reference to the term in question. Instead, it takes you to the first mention of the term in the text, plus any pages featuring substantial further work on the area. On these pages you will normally find an explanation and some illustrative examples.